Introduction to the Mechanics of Deformable Solids

Reproduction of a sketch of a cantilevered beam taken from Galileo Galilei's textbook *Discourses on Two New Sciences* (1638), {{PD-1923}}.

David H. Allen

Introduction to the Mechanics of Deformable Solids

Bars and Beams

 Springer

David H. Allen
College of Engineering and Computer Science
University of Texas-Pan American
Edinburg, TX, USA

ISBN 978-1-4614-4002-4   ISBN 978-1-4614-4003-1 (eBook)
DOI 10.1007/978-1-4614-4003-1
Springer New York Heidelberg Dordrecht London

Library of Congress Control Number: 2012937796

Printed on acid-free paper

Springer is part of Springer Science+Business Media (www.springer.com)

To Claudia, who understands what it means to be married to an engineer, and nevertheless is.

Preface

Archeologists tell us that sometime around 13,000 years ago, in the then fertile valleys of the Tigris and Euphrates Rivers, several enterprising individuals began to divert water from the rivers for the purpose of growing crops. That may indeed have been the single most important step in creating our modern civilization today, and for that reason it is not uncommon to refer to those ingenious people as the first engineers on earth. Concomitantly, anthropologists tell us that it takes approximately 10,000 years for the human species to undergo any significant genetic change. Thus, we may conclude that those people were very smart, perhaps as smart as we are today, as indicated by evidence that their brains are essentially the same size as that of modern man and woman. Perhaps antithetically, the birds and the bees, the frogs and the trees, indeed virtually all other species of plants and animals on earth have undergone only modest evolution in the past 13,000 years. Yet our own species has in that (geologically) short span of time taken over this planet.

So what is going on here? Why have we humans changed so dramatically, while other species have not? The answer is of course—education. Our species is the first species, so far as we are aware, that has outrun our own evolution, and we have done so via education. Certainly Darwin's law played a great role in our quest to educate ourselves, despite the fact that it was not even espoused until the mid-nineteenth century. The fact is, our ancestors were living their lives according to Darwin's law, whether they were aware of it or not. Archeologists have determined that the invention of farming moved humans rapidly away from hunter-gatherer behaviors, and this produced a population explosion at places such as Ur in the Middle East. Apparently, within a few short centuries, cities of more than 10,000 persons had sprung up in the Mideast. The growth of these cities allowed for specialization of professions in these cities, and this led the way to the development of new technologies as more people specialized in the development of new ideas. In turn, the development of new ideas required some training, and the ability to transmit these developments through society necessitated the development of sophisticated mathematics and language. These developments led inexorably to the rise of education—a necessity for humans to survive. While the higher education complexes on our planet are essentially less than a 1,000 years old, our educational infrastructure goes back thousands of years. It may be argued that education is indeed the single most important development in the history of humankind.

I was born into the world of blackboards and chalk, slide rules, and hand-drawn graphs. Now, as I near the twilight of my career in academia, I find myself to be a euphemism for the dinosaurs of old. The way that I was taught when I was in school half a century ago is no longer germane to our society. The tools that we used are virtually all obsolete, and here is the most astonishing part—*this is the first time in recorded history that our technology has outrun our education in a single life span.* And yet, here I sit, working at a university, attempting to educate people less than a third my age—people who grew up in a world that I did not—people who are comfortable with cell phones, ipods, ipads, 3-D television, GPS, and I could go on and on. But more importantly, people who are **NOT** comfortable with slide rules, trig tables, analytic geometry, rigorous analytic methods for solving differential equations, and hand-drawn graphs. Are these people ill-educated? Are they not prepared for college? These are questions that are beyond the scope of this textbook. But what I do know is that they are different—they are different from my generation. They think differently, and *they learn differently.*

I have been teaching for almost 40 years. I remember when I was in college one of my professors (Dr. Thompson was his name) came to class the first day and announced, "I have been teaching 40 years. I have taught every way there is to teach, and all of them are wrong." That statement has stuck with me these past 40 years, and now I find myself on the other end of the problem. You see, I feel the same way he did. And because the evolution of technology has increased its pace, I fear that Dr. Thompson's conjecture is even more relevant today.

I have been teaching subject matter related to the subject of this textbook for my entire professional career. When I surveyed the available textbooks on this subject recently, I was surprised to find that while technology has changed, while America's youth have clearly changed, the approach taken to teaching this subject has not materially changed in the past 40 years. Actually, if one studies the mid-twentieth century texts by S.P. Timoshenko and his colleagues, it will be apparent that little has changed in significantly more than 40 years.

To my dismay, I found the following revelations within the subject matter that I reviewed: little attention to mathematical rigor, little or no attention to the pursuit of fundamental knowledge, wholesale attention to trivial details, and poor attention to ultimate outcome—*understanding of the subject*. While it is true that mechanics is a very old discipline, perhaps even the oldest of scientific disciplines, it is nonetheless clear that much has changed in the field of mechanics over the past half century. A great deal of this change has come about due to the birth and growth of the computer age. Armed with Moore's law, mechanicians have dramatically changed and improved our field of engineering and science. Especially in the field of deformable body mechanics, the inexorable spread of the finite element method over the past half century has revolutionized our ability to model deformable bodies today. And yet, we seem to have failed to alter our educational approach to the subject.

This textbook is an attempt to address this problem—to approach a first course in deformable body mechanics in such a way as to impart fundamentals to the student that will lead the student to a rigorous and logical understanding of the field as it is utilized today—in the world of high speed computing. As such, it is intended that students who master the subject matter herein will find within their grasp the ability to progress seamlessly to a second course wherein they will learn to design real world complicated and three-dimensional structural parts using already available software.

My approach herein grew out of my 40-year career in higher education, during which time I taught at four different major universities in the USA—Texas A&M University, Virginia Tech University, The University of Nebraska-Lincoln, and The University of Texas-Pan American. Over that span of time I taught courses such as statics, dynamics, mechanics of solids, advanced mechanics of solids, finite element methods, advanced structural mechanics, elasticity, plasticity, viscoelasticity, viscoplasticity, fracture mechanics, and the history of science and technology to more than 6,000 ensemble students. Perhaps serendipitously, toward the middle of my career, I taught within an experimental course sequence funded by the National Science Foundation for 13 years. This period profoundly affected my thinking on the subject.

My aim herein is to impart fundamentals with as little confusion as possible. For example, I have adopted a systematic mathematical terminology, taken at least in part from my previous textbook on the subject *Introduction to Aerospace Structural Analysis* coauthored with Walter E. Haisler. Furthermore, my intention is for the student who masters the subject matter herein to be competent to move directly to a course wherein the mechanics of deformable bodies can be modeled either two or fully three dimensionally using the finite element method. Thus, I have purposely avoided many topics that the interested reader can find in the enormous body of texts dealing with the subject of mechanics of deformable bodies.

The text opens with a short history of mechanics. This chapter is by no means exhaustive on the subject, aiming to impart the high points of historical developments that have led to our modern day understanding. The second chapter of the book deals with the underpinnings of our present day models, including fundamental universal conservation laws, definitions of the essential variables in the model, such as stress and strain, and a brief introduction to constitutive behavior of deformable bodies.

The third chapter develops the theory of uniaxial bars. Interestingly, this theory seems to have been developed after the theory of beams was developed by Leonard Euler and Daniel Bernoulli in the mid-eighteenth century, despite the fact that beams are far more physically and mathematically complicated. Perhaps it was expedience that drove Euler and Bernoulli to address the beam problem first. After all, beams were and still are our most important structural elements, whereas uniaxial bars are less prominent and perhaps more significantly, less prone to failure. Nonetheless, the chapter on uniaxial bars is of great importance for the student who is just starting out in this subject for two reasons: (1) it will pave the way toward an intimate understanding of the more complicated theories of torsion and beams and (2) the theory of uniaxial bars contains all of the essential physics of the general three-dimensional theory of elasticity employed in finite element algorithms without the encumbrance of complicated mathematics such as partial differential equations.

The fourth chapter develops the theory of torsion bars. I personally enjoy this subject because of the mathematical similarity of the torsion theory to uniaxial bar theory, despite the totally different physics involved. Such congruencies occur often in nature in widely differing fields of study, thus forming a bridge for those who are drawn to change their discipline. This chapter also forms a nice connection between the rather straightforward subject of uniaxial bars and the more challenging subject of beams.

The fifth chapter develops the theory of what I call "simple" beams. By simple I mean (1) beams that do not undergo axial extension and bending simultaneously, (2) beams whose properties vary only in their long direction, (3) beams that are initially straight, (4) beams that undergo small deformations, (5) beams that are orthotropic and linear elastic, and (6) beams that are not subjected to temperature change. As confining as these restrictions are, the theory developed in this chapter is nevertheless powerful for many practical applications. More importantly, the theory encompasses essentially all of the necessary knowledge for understanding the mechanical behavior of beams. For the reader who is interested in more advanced beams, I refer you to my previous textbook, cited above.

The sixth chapter of the book discusses the aspects of analyzing slender structural components. These include (1) the introduction of the principle of superposition and how it may be used as a practical simplifying tool, (2) the subject of stress transformations (due to coordinate rotations), and (3) how a rigorous understanding of this important but complicated subject is essential for the purpose of determining whether structural components can be expected to fail due to yielding and/or fracture.

The seventh and final chapter of the book briefly introduces the subject of structural design. While the subject of design is often quite open-ended and artistic in nature, the approach taken herein is simplistic in the sense that design is viewed as an inverse problem wherein the typical outputs that result from structural analysis of a part with a priori chosen loads, geometry, and material properties is inverted to a form in which these inputs become the outputs. A successful design will be considered to be any choice of loads, geometry, and material properties that satisfies all of the design constraints. No attempt will be made to produce an optimized design, as this constitutes an advanced subject that is beyond the scope of the present text. Rather, the goal of this closing chapter in the current text will be to explore in a straightforward manner the power of the models developed in Chaps. 1–6 of this text.

My experience is that essentially all of the material contained in this textbook can be covered in a single semester to typical university students in the USA.

Edinburg, TX, USA   David H. Allen

Acknowledgments

I would like to thank all those who helped to make this textbook come to fruition. I am especially indebted to my editor at Springer, Michael Luby, and his assistant Merry Stuber. Special thanks are due to Dr. George Gazonas, who read the manuscript and made valuable suggestions that markedly improved the final text.

Contents

Photo Credits

Chapter 1

Fig. 1.1 Painting of Archimedes by Domenico Fetti (1620), reproduced on Creative Commons public domain {{PD-Art}}, accessed in March 2012 at http://en.wikipedia.org/wiki/File:Domenico-Fetti_Archimedes_1620.jpg

Fig. 1.3a Typical page from the Archimedes Palimpsest, source http://www.archimedespalimpsest.net, approved for reproduction by Walters Museum Curator Walter Noel under Creative Commons License (CC BY 3.0)

Fig. 1.3b Typical page after imaging from the Archimedes Palimpsest, source http://www.archimedespalimpsest.net, approved for reproduction by Walters Museum Curator Walter Noel under Creative Commons License (CC BY 3.0)

Fig. 1.4 Stamp Issued in the Soviet Union in 1983 to Commemorate the 1,200th Birthday of Al-Kwarizmi. This work is not an object of copyright according to Part IV of Civil Code No. 230-FZ of the Russian Federation of December 18, 2006 {{PD-1923}}, accessed on Creative Commons in March 2012 at http://en.wikipedia.org/wiki/File:Abu_Abdullah_Muhammad_bin_Musa_al-Khwarizmi_edit.jpg

Fig. 1.8 Self-Portrait of Leonardo Da Vinci, c. 1512 Biblioteca Real, Turin, reproduced on Creative Commons public domain {{PD-Art}}, accessed in March 2012 at http://en.wikipedia.org/wiki/File:Leonardo_self.jpg

Fig. 1.10 Portrait of Nicolaus Copernicus from Thorn Town Hall, 1580, reproduced on Creative Commons public domain {{PD-Art}}, accessed in March 2012 at http://en.wikipedia.org/wiki/File:Nikolaus_Kopernikus.jpg

Fig. 1.11 Portrait of Tycho Brahe by Edouard Ender from his book *Astronomiae Instauratae Mechanica*, 1598, reproduced on Creative Commons public domain {{PD-Art}}, accessed in March 2012 at http://en.wikimedia.org/wiki/File:Tycho_Brahe.jpg

Fig. 1.12 Portrait of Johannes Kepler by an unknown artist, 1610, reproduced on Creative Commons public domain {{PD-Art}}, accessed in March 2012 at http://en.wikipedia.org/wiki/File:Johannes_Kepler_1610.jpg

Fig. 1.13 Portrait of Galileo Galilei in 1636 by Justus Sustermans, reproduced on Creative Commons public domain {{PD-Art}}, accessed in March 2012 at http://en.wikipedia.org/wiki/File:Justus_Sustermans_-_Portrait_of_Galileo_Galilei,_1636.jpg

Fig. 1.15 Image taken from Galileo's *Dialogues Concerning Two New Sciences* (Galilei 1638), {{PD-1923}}

Fig. 1.17 Portrait of René Descartes by Frans Hals in the Louvre Museum, reproduced on Creative Commons public domain {{PD-Art}}, accessed in March 2012 at http://en.wikipedia.org/wiki/File:Frans_Hals_-_Portret_van_Ren%C3%A9_Descartes.jpg

Fig. 1.18 Plate to Robert Hooke's Lecture "Of Spring," 1678, {{PD-1923}}

Fig. 1.19 Portrait of Isaac Newton by Sir Godfrey Kneller, reproduced on Creative Commons public domain {{PD-Art}}, accessed in March 2012 at http://en.wikipedia.org/wiki/File:GodfreyKneller-IsaacNewton-1689.jpg

Fig. 1.20a Portrait of Jacob Bernoulli by an unknown author, reproduced on Creative Commons public domain {{PD-Art}}, accessed in March 2012 at http://en.wikipedia.org/wiki/File:Jakob_Bernoulli.jpg

Fig. 1.20b Portrait of Johann Bernoulli by an unknown author, reproduced on Creative Commons public domain {{PD-Art}}, accessed in March 2012 at http://en.wikipedia.org/wiki/File:Johann_Bernoulli2.jpg

Fig. 1.21 Portrait of Daniel Bernoulli by Johann Jacob Haid, reproduced on Creative Commons public domain {{PD-Art}}, accessed in March 2012 at http://en.wikipedia.org/wiki/File:Daniel_Bernoulli_001.jpg

Fig. 1.22 Portrait of Leonhard Euler by Johann Georg Brucker, reproduced on Creative Commons public domain {{PD-Art}}, accessed in March 2012 at http://en.wikipedia.org/wiki/File:Leonhard_Euler_2.jpg

Fig. 1.23 Portrait of Joseph-Louis Lagrange by Giuseppe Lodovico, reproduced on Creative Commons public domain {{PD-Art}}, accessed in March 2012 at http://en.wikimedia.org/wiki/File:Langrange_portrait.jpg

Fig. 1.24 Posthumous Portrait of Pierre-Simon Laplace by Madame Feytaud (1842), reproduced on Creative Commons public domain {{PD-Art}}, accessed in March 2012 at http://en.wikimedia.org/wiki/File:Pierre-Simon_Laplace.jpg

Fig. 1.25 Portrait of Ernst Chladni, reproduced on Creative Commons public domain {{PD-Art}}, accessed in March 2012 at http://en.wikipedia.org/wiki/File:Echladni.jpg

Fig. 1.28 Portrait of Siméon-Denis Poisson, reproduced on Creative Commons public domain {{PD-Art}}, accessed in March 2012 at http://en.wikipedia.org/wiki/File:Simeon_Poisson.jpg

Fig. 1.29 Bust of Claude-Louis Navier, reproduced on Creative Commons public domain {{PD-Art}}, accessed in March 2012 at http://en.wikipedia.org/wiki/File:Claude-Louis_Navier.jpg

Fig. 1.30 Sketch of Joseph Fourier by an unknown artist c. 1820, reproduced on Creative Commons public domain {{PD-Art}}, accessed in March 2012 at http://en.wikipedia.org/wiki/File:Joseph_Fourier_(circa_1820).jpg

Fig. 1.31 Photo from Smithsonian Institution Libraries of Augustin Cauchy taken circa 1856 by E.H. Reutlinger, reproduced on Creative Commons public

domain {{PD-US}}, accessed in March 2012 at http://en.wikipedia.org/wiki/File:
Augustin-Louis_Cauchy.jpg

Fig. 1.32 Photograph of Gabriel Lamé, reproduced on Creative Commons public domain {{PD-Art}}, accessed in March 2012 at http://en.wikipedia.org/wiki/File:
Gabriel-Lam%C3%A9.jpeg

Fig. 1.33 Painting of Christian Otto Mohr by Osmar Schindler, reproduced on Creative Commons public domain {{PD-Art}}, accessed in March 2012 at http://en.wikipedia.org/wiki/File:Otto_Mohr.JPG

Fig. 1.34 Portrait of Thomas Young in the 1820s by Sir Thomas Lawrence, reproduced on Creative Commons public domain {{PD-Art}}, accessed in March 2012 at http://en.wikipedia.org/wiki/File:Young_Thomas_Lawrence.jpg

Chapter 2

Fig. 2.3 Photo of Canopus at the Villa Adriana taken by Wikipedia contributor Andy Hay and published under Creative Commons License (CC BY 2.0), accessed in March 2012 at http://upload.wikimedia.org/wikipedia/commons/thumb/6/64/
Villa_Adriana.jpg/800px-Villa_Adriana.jpg

Figure from Problem 2.2 Photo of Hoover Dam Releasing Water (courtesy Bureau of Reclamation PD-USGOV)

Chapter 4

Fig. 4.2a Photo of US Army AH 64-D Apache Longbow helicopter, source, http://
www.army.mil (courtesy US Army)

Fig. 4.2b Photo of vertical axis wind turbine in Gaspesie, Quebec, Canada, photo taken by Wikipedia contributors Joanne and Matt and published under Creative Commons License (CC BY 2.0), accessed in March, 2012 at http://commons.
wikimedia.org/wiki/File%C3%89olienne_verticale.jpg

Chapter 6

Fig. 6.2 Photo of Firth of Forth Bridge taken by Wikipedia contributor George Gastin and published under Creative Commons License (CC BY 3.0), accessed in March 2012 at http://en.wikipedia.org/wiki/File:Forth_bridge_evening_long_
exposure.jpg

Fig. 6.3 Photo of Brooklyn Bridge taken by Wikipedia contributor Peter Zoon and published under Creative Commons License (CC BY 2.0), accessed in March 2012 at http://commons.wikimedia.org/wiki/File:Brooklyn_Bridge_-_Flickr_-_Peter_Zoon.jpg

Fig. 6.4a Photo of Sydney Harbor Bridge taken by Wikipedia contributor Adam J.W.C. and published under Creative Commons License (CC BY 3.0), accessed in March 2012 at http://de.wikipedia.org/w/index.php?title=Datei:
Sydney_harbour_bridge_new_south_wales.jpg&filetimestamp=2010078015222

Fig. 6.4b Photo of Golden Gate Bridge taken by Wikipedia contributor Rich Niewiroski Jr. and published under Creative Commons License (CC BY 2.5), accessed in March 2012 at http://en.wikipedia.org/wiki/File:GoldenGateBridge-001.jpg

Fig. 6.5 Photo of RMS Titanic departing Southampton on April 10, 1912, photo by F.G.O. Stuart, accessed on Creative Commons public domain {{PD-1923}}in March 2012 at http://en.wikipedia.org/wiki/File:RMS_Titanic_3.jpg

Fig. 6.6 Photos of the Challenger Disaster; Explosion on the left; Challenger underwater on the right (photos courtesy NASA)

Fig. 6.7 Photo of full scale model of section of Chunnel at National Railway Museum in York, England uploaded by Wikipedia contributor Xtrememachineuk under Creative Commons License (CC BY 2.5), accessed in March 2012 at http://en.wikipedia.org/wiki/File:Channel_Tunnel_NRM.jpg

Fig. 6.8 Aerial Photo of Charles De Gaulle Airport Terminal 1 taken by Wikipedia contributor citizen59 and published under Creative Commons License (CC BY 3.0), accessed in March 2012 at http://commons.wikimedia.org/wiki/File:Terminal1_CDGParisAirport.jpg

Fig. 6.9 Photo of Aloha Airlines Disaster, photo courtesy FAA

Fig. 6.10a Photo of bow of Wasa taken by Wikipedia contributor Nick Lott and published under Creative Commons License (CC BY 2.0), accessed in March 2012 at http://commons.wikimedia.org/wiki/File:The_Wasa_from_the_Bow.jpg

Fig. 6.10b Photo of stern of Wasa taken by Wikipedia contributor Nick Lott and published under Creative Commons License (CC BY 2.0), accessed in March 2012 at http://commons.wikimedia.org/wiki/File:The_Wasa_from_the_stern.jpg

Fig. 6.12 Photo of reconstruction of the stage 1 fan disk in the Sioux City Aircraft Crash of flight UA 232 on July 19, 1989 (photo courtesy NTSB, 1990)

Fig. 6.13 Photo on left of the Minneapolis Bridge Collapse (courtesy US Coast Guard); Photo on right showing fracture in gusset plate (courtesy NTSB)

Fig. 6.14a Photo of Boeing 787 Dreamliner taken by Wikipedia contributor 787_First_Flight.jpg:Dave Sizer and published under Creative commons License (CC BY 2.0), accessed in March 2012 at http://en.wikipedia.org/wiki/File:Boeing_787_first_flight.jpg

Fig. 6.14b Photo of Airbus A380 taken by Wikipedia contributor Axwel and published under Creative Commons License (CC BY 2.0), accessed in March 2012 at http://en.wikipedia.org/wiki/File:Airbus_A380_blue_sky.jpg

Fig. 6.26 Lines of Principal Planes (left) in a Curved Crane Compared to the Trabecular Alignment in a Human Proximal Femur (right) photo published in (Thompson 1917) from previous publications by (Culmann 1866, Meyer 1867, Wolff 1870), photo in the public domain under {{PD-US}}

Chapter 7

Fig. 7.2 Photo of Meidoum Pyramid taken by Wikipedia contributor Neithsabes and released into the public domain, accessed in March 2012 at http://en.wikipedia.org/wiki/File:Meidoum_pyramide_003.JPG

Fig. 7.4 Photo of Red Pyramid uploaded to en.wikipedia by Ivrienen and transferred to Commons by Wikipedia contributor User:GDK under Creative Commons License (CC BY 3.0), accessed in March 2012 at http://en.wikipedia.org/wiki/File:Snofrus_Red_Pyramid_in_Dahshur_(2).jpg

Fig. 7.5a Photo of Amiens Cathedral taken by Wikipedia contributor Florestan and published under Creative Commons License (CC BY 3.0), accessed in March 2012 at http://commons.wikimedia.org/wiki/File:Facade_de_la_cathedrale_d%27Amiens.jpg

Fig. 7.8 Artist's Rendition of the National Aerospace Plane (courtesy NASA)

Fig. 7.11 C5A Galaxy Military Aircraft (photo courtesy USAF)

Figure from Problem 7.4 Photo of Norwegian gymnast Espen Jansen performing at the Norwegian National Championships in 2001. Photo taken by Wikipedia contributor Stian Fugelsøy and published under Creative Commons License (CC BY 3.0), accessed in March 2012 at http://commons.wikimedia.org/wiki/File:Espen_Jansen_-_NMturn15.jpg

Chapter 1
A Short History of Mechanics

1.1 Historical Review

Mechanics is the study of the motion of bodies. Because our ancestors have long been interested in the motions of heavenly bodies, mechanics is perhaps the oldest of the sciences. In the words of George Santayana (1863–1952), "Those who cannot remember the past are condemned to repeat it" (Santayana 1905). Mindful of this pithy but ominous remark, it is perhaps advisable to recount some of the more important developments that led the way to the modern theories for deformable bodies that are widely in use today. However, as an exhaustive coverage of this subject is beyond the scope of this text, this review concentrates on the more significant advances that have contributed to the development of models for deformable bodies. To those whose mentionable works have not been expounded here, the author apologizes and points toward the culprit "expedience."

Certainly there were many Greek scientists before and during the Hellenistic period who studied the motion of bodies, and while Aristotle (384–322 BC) is known to have expounded the principle of the lever, the historical record must surely point to Archimedes (287–212 BC) as the most important mechanist from antiquity, shown in Fig. 1.1. Archimedes gave a detailed account of the principles associated with the lever, and although slightly flawed, they can be said to contain the essential components of modern statics. They may also be viewed as a forerunner of the principle of virtual work.

While Archimedes' achievements with the lever alone would certainly ensure his place in the history of mechanics, there was much more to come from this great scientist. He also expounded the principle of buoyancy in great detail, thus recording the first significant results on deformable bodies and their properties.

Archimedes is also known to have proven the relationship between the circumference of a circle and the area, in the process estimating the value of pi quite accurately. Furthermore, he is known to have produced a device for measuring the movements of the known planets, the sun, and the moon—an astronomical clock if

D.H. Allen, *Introduction to the Mechanics of Deformable Solids: Bars and Beams*, DOI 10.1007/978-1-4614-4003-1_1, © Springer Science+Business Media New York 2013

Fig. 1.1 Painting
of Archimedes
by Domenico Fetti (1620)

you will. A device called the Antikythera clock was discovered in the Mediterranean in 1901, and it is believed to be a copy (or perhaps the original?) of the device constructed by Archimedes more than 2,000 years ago, as shown in Fig. 1.2.

But there is still more that came from the mind of Archimedes. In his apparent quest to understand the buoyancy of ships, he produced perhaps the most remarkable theorem from antiquity, calculating the center of gravity of a parabolic cylinder. In so doing, he used the method of exhaustion [attributed to Eudoxus (410–355 BC)], in such a way as to introduce the concept of infinity, thereby pointing the way toward the theory of modern calculus, which is a necessary component in any cogent theory of mechanics.

Interestingly, the last of the above developments attributed to Archimedes, although referred to in other literary sources, was not fully verified until a palimpsest was sold at auction at Christie's in 1998, and is now on exhibit at the Walters Art Museum in Baltimore, Maryland (Netz and Noel 2009), as shown in Fig. 1.3. The proof, found in this palimpsest, has verified the importance of Archimedes to the history of mathematics as well as mechanics. Although the above discoveries by Archimedes are by no means his only scientific contributions, they are more than sufficient to rank him as the greatest of the ancient mechanists.

Fig. 1.2 Front and rear view photo of the Antikythera clock in the Athens Museum of antiquities

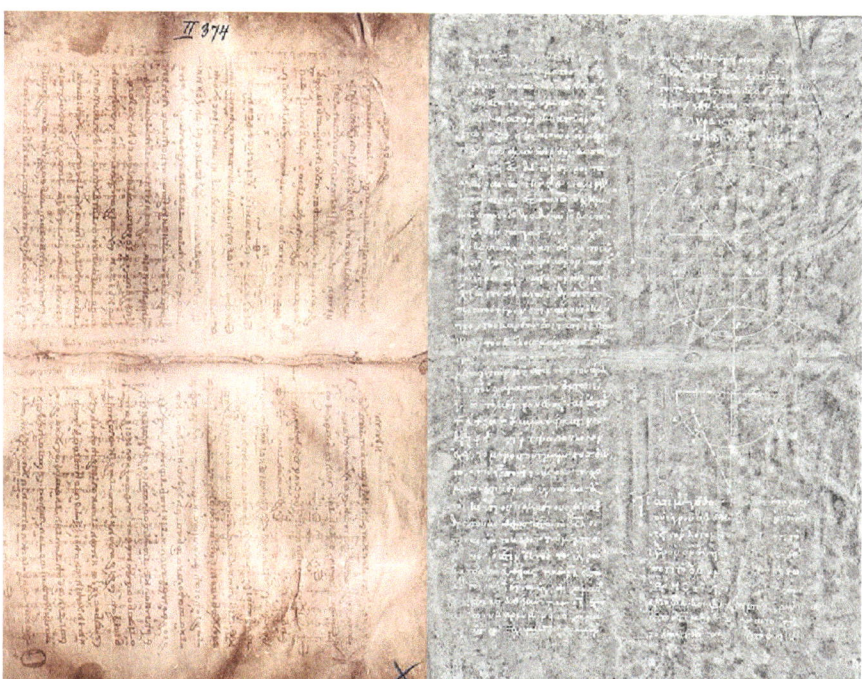

Fig. 1.3 A page from the Archimedes Palimpsest; photo of the overwritten text on the *left*, image processed photo on the *right* showing Archimedes' text beneath

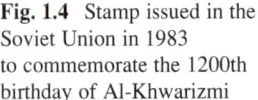

Fig. 1.4 Stamp issued in the
Soviet Union in 1983
to commemorate the 1200th
birthday of Al-Khwarizmi

It is now known that significant advances in science occurred in other cultures
during ancient times, principally in Mesopotamia, Egypt, India, and China
(Teresi 2003). However, historians today are still determining how much of
an impact these developments had on western culture. Suffice it to say that the
ancient Greeks were undoubtedly influenced profoundly by other cultures both
before and during their time.

With the fall of the Roman Empire in 476 the written language was largely lost
for nearly 1,000 years in most of Western Europe. The lone exception seems to have
been in Southern Spain, which was occupied by Middle Eastern cultures until the
Renaissance. Scientists from these cultures retained much of the extant documents
from the ancients that we have today. They also made significant advances along
the way, such as the incorporation of Hindu numbers into Western Culture, which
came down to us from the ninth century Persian mathematician Al-Khwarizmi
(c780-c850) (Fig. 1.4). His text introduced al-jabr (algebra) and was translated by
Leonardo of Pisa, also known as Fibonacci, in the late twelfth century (Fig. 1.5).
The Hindu numerals used in Al-Khwarizmi's book, and employed by Fibonacci, are
essentially the modern numbers that we use today.

The Renaissance began as an artistic movement in the early fourteenth century,
but evolved into a technological rebirth that necessitated the development of
new scientific tools. One surely singular event toward the middle of the Renaissance that hastened the rebirth of science was the construction of the dome
of the Santa Maria del Fiori in Florence (see Fig. 1.6) in 1420–1434 by Filippo

Fig. 1.5 Statue of Leonardo
of Pisa (Fibonacci)
in the Camposanto in Pisa

Brunelleschi (1377–1446) (King 2001) depicted in Fig. 1.7. His innate understanding of the principle of the dome led to its construction without the necessity to build scaffolding, thus foreshadowing the race to create robust structural models. The development of mathematical models for elastic bodies almost 500 years later would confirm Brunelleschi's prescience.

Toward the end of that same century Leonardo Da Vinci (1452–1519), ever the dabbler, recorded what may be the first systematic attempt in history to measure the strength of a material. Leonardo, shown in Fig.1.8, writes in one of his manuscripts (circa 1500):

> The object of this test is to find the load an iron wire can carry. Attach an iron wire 2 bracchia (about 1.3 m) long to something that will firmly support it, then attach a basket or any similar container to the wire and feed into the basket some fine sand through a small hole placed at the end of a hopper. A spring is fixed so that it will close the hole as soon as the wire breaks. The basket is not upset while falling, since it falls through a very short distance. The weight of sand and the location of the fracture of the wire are to be recorded. The test is repeated several times to check results. Then a wire of one-half the previous length is tested and the additional weight it carries is recorded, then a wire of one-fourth length is tested and so forth, noting each time the ultimate strength and the location of the fracture.

(Reti 1990)

Fig. 1.6 Photo of the Brunelleschi dome in Florence

Fig. 1.7 Bust of Brunelleschi in the Santa Maria Del Fiori

Fig. 1.8 Self-Portrait
of Leonardo Da Vinci,
c. 1512 Biblioteca Real, Turin

It is also known today that Da Vinci investigated the behavior of beams, as recorded in the Madrid Codex I (Reti 1990). In this manuscript, Da Vinci clearly anticipated the Euler–Bernoulli assumption by more than a century. Unfortunately, it is not yet clear whether Leonardo's views on the subject were made public.

Sadly, the efforts undertaken by Brunelleschi, Da Vinci (see his last home in Fig. 1.9), and others during the Renaissance did not immediately lead to significant advances in mechanics models, as most sages of that time period clung to the old Aristotelian and Ptolemaic principles.

Nicolaus Copernicus (1473–1543), shown in Fig. 1.10, was perhaps the first person in the modern era to come forward and espouse a (new) theory of great significance. His *De Revolutionibus Orbium Coelestium* (Copernicus 1543) is considered by many to be the most important scientific book ever written. In it he renounced the Ptolemaic theory of the earth as the center of the universe and instead placed the Sun at the center. It is now known that Copernicus' book was read by many great scientists who came thereafter.

More than half a century passed before additional significant advances would occur in mechanics, and the initial ones would come principally from three great scientists: Tycho Brahe (1546–1601) (Fig. 1.11), Johannes Kepler (1571–1630)

Fig. 1.9 Photo of Leonardo Da Vinci's last home, the Clos Lucé, in Amboise, France

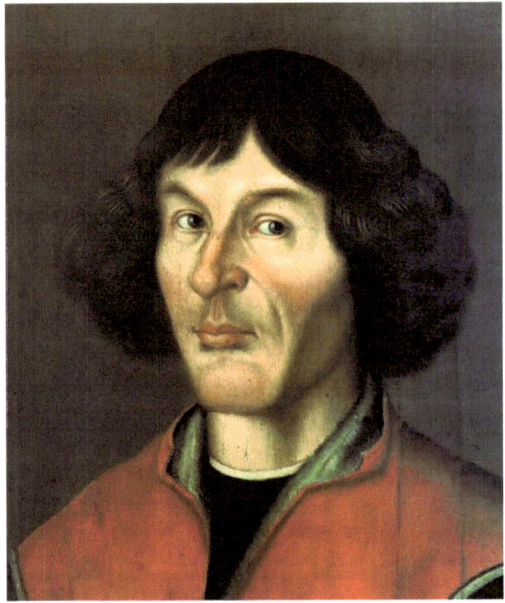

Fig. 1.10 Portrait of Nicolaus Copernicus from Thorn Town Hall, 1580

(Fig. 1.12), and Galileo Galilei (1564–1642) (Fig. 1.13). Brahe was perhaps the first great modern astronomer, patiently recording the movements of the heavenly bodies over the span of a lifetime without the benefit of the telescope, which would come a short time later. Kepler, who became Brahe's assistant, correctly

Fig. 1.11 Portrait of Tycho Brahe by Edouard Ender from his book Astronomiae instauratae mechanica, 1598

Fig. 1.12 Portrait of Johannes Kepler by an unknown artist, 1610

Fig. 1.13 Portrait of Galileo Galilei in 1636 by Justus Sustermans

interpreted Brahe's measurements, and created a model for predicting the motions of the planets that is still accurate today.

The third of this triumvirate, Galileo, was destined to become one of the half dozen most important scientists in the history of humankind. Not only did Galileo espouse the law of falling bodies (a conjugate to Kepler's laws of the motions of heavenly bodies), he would in the course of perfecting the scientific telescope (see Fig. 1.14) discover the moons of Jupiter and the mountains of the moon. These discoveries were sufficient to justify his contentions that the old Aristotelian and Ptolemaic beliefs were in error and that Copernicus was correct. Unfortunately, his discoveries soon brought Galileo into conflict with the Catholic Church. This ultimately led to his conviction by the Inquisition for heresy, occurring at a time when burning at the stake was not an unlikely sentence. However, because he publically recanted his conviction that the Ptolemaic (earth centered) universal system was incorrect, he was allowed to spend the last 8 years of his life under house arrest at his villa at Arcetri, on the hill overlooking Florence. There he died blind, incontinent, and alone.

While Galileo was not vindicated during his lifetime, his enormous legacy is recognized today. Toward the end of his life, he managed to smuggle his most profound manuscript to Germany, thus defining the modern science of mechanics in his seminal work *Discourses and Mathematical Demonstrations Relating to Two New Sciences* (Galilei 1638) (Fig. 1.15). More importantly, Galileo is today given

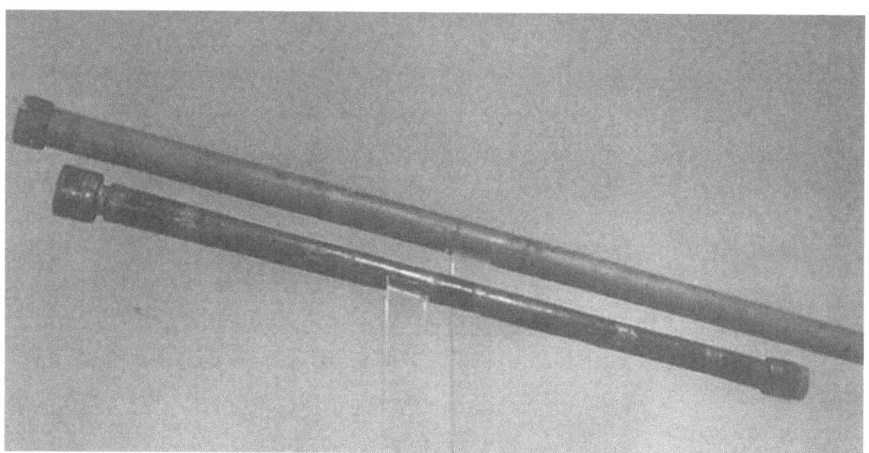

Fig. 1.14 Photo of Galileo's Telescopes in the Galileo Museum in Florence

Fig. 1.15 Image taken from Galileo's *Dialogues Concerning Two New Sciences* (Galilei 1638)

Fig. 1.16 Photo of Galileo's tomb in the Santa Croce Basilica, Florence

much of the credit for establishing the foundations of the modern scientific method. His magnificent tomb, though he was not accorded this place of honor for almost a century after his death, is today in the Santa Croce Cathedral in Florence, as shown in Fig. 1.16. Indeed, according to Albert Einstein, Galileo is "the father of modern physics—indeed of modern science altogether" (Einstein and Calaprice 1996).

A contemporary of Galileo, René Descartes (1596–1650) (Fig. 1.17) led the French school in the early sixteenth century, though he spent much of his life in other countries, including Holland. Although it appears that Pierre de Fermat (1601–1665) was the first to employ a three-dimensional rectilinear coordinate system, Descartes utilized the coordinate system that bears his name (Cartesian coordinates) in his exhaustive textbook *La Géométrie* (Descartes 1954). While Descartes is not remembered so much for contributions to mechanics, he is one of the first great philosophers of modern times. Furthermore, his deployment of algebra in preference to geometry was revolutionary for his time and helped to pave the way for many of the developments to come in mechanics.

Toward the middle of the seventeenth century Robert Hooke (1635–1703) began performing experiments on all manner of metallic springs (see Fig. 1.18). His results, "ut tensio sic vis" (Jardine 2005), became known as Hooke's Law, the forerunner of linear elastic constitutive models that are necessary to predict the mechanical response of linear elastic bodies.

Isaac Newton (1642–1727) (Fig. 1.19) was born the year that Galileo died. He was to become perhaps the greatest scientist who ever lived. His oeuvre includes

Fig. 1.17 Portrait of René Descartes by Frans Hals in the Louvre Museum

no less than the establishment of the universal law of gravitation, the universal laws of motion, and the invention of calculus [jointly with Gottfried Leibniz (1646–1716)]. His contributions to our world are so important as to pervade virtually every aspect of our lives. His laws and his mathematics form the corner-stone of modern mechanics. So enormous was his impact that within a few short months of the publication of his monumental book *Philosophiae Naturalis Principia Mathematica* (Newton 1686), he had become perhaps the most famous person on earth. Such a lofty status has only been accorded a scientist a scant few times in the history of mankind.

The theories enunciated by Newton were to be examined and verified in the early eighteenth century, especially by Christiaan Huygens (1629–1695). But whereas initial efforts were preoccupied with the study of so-called rigid bodies (such as the planets), toward the middle half of the century scientists began to turn their attentions toward the development of accurate models for predicting the mechanics of so-called deformable bodies. History records that the first attempts to use calculus to model deformable bodies were due to two of the Bernoulli's from Basel, brothers Jacob (1654–1705) and Johann (1667–1748) (Fig. 1.20). Jacob proposed that the curvature of a beam is proportional to the bending moment. Johann introduced the principle of virtual displacements. Although their attempts were incomplete, they paved the way for the first useful theory of beams.

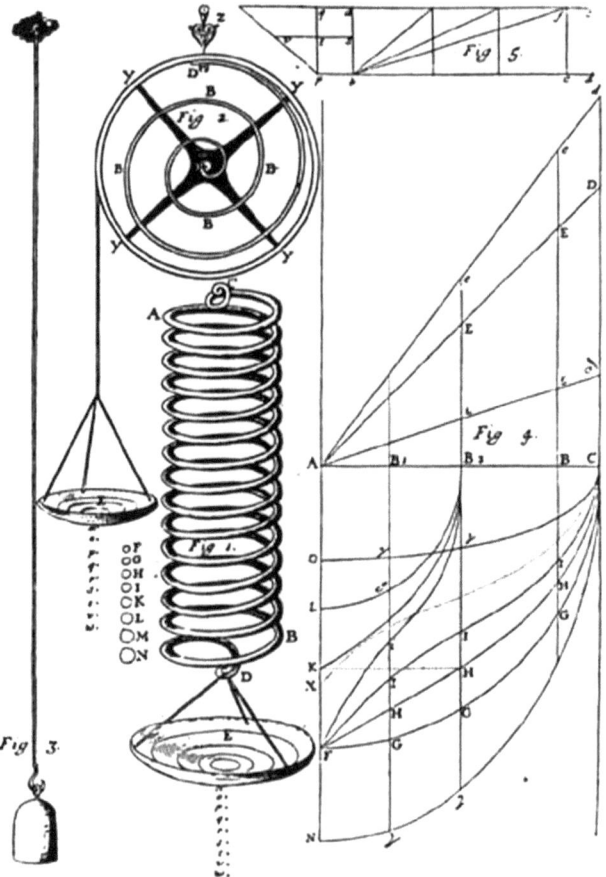

Fig. 1.18 Plate to Robert Hooke's Lecture "of Spring," 1678

Johann Bernoulli's son Daniel Bernoulli (1700–1782) (Fig. 1.21) became the chair of mathematics at The Russian Academy of Sciences in St. Petersburg in 1725. Shortly thereafter his former pupil from Basel, Switzerland, Leonhard Euler (1707–1783) (Fig. 1.22), joined him in St. Petersburg. Daniel would leave St. Petersburg in 1733, but he continued to correspond with Euler. Sometime during the succeeding years Daniel proposed in a letter a method of solution for beams that Euler took up in earnest, and this model appeared in Euler's book *Methodus Inveniendi Lineas Curvas* (Euler 1744). This approach is still in use today for designing structural components. The "Euler–Bernoulli beam theory", as it is now called, is one of the seminal events in the development of theories for modeling the deformations of solids.

Euler was one of the most important scientists of the eighteenth century, establishing the first complete three-dimensional mathematical models of the

Fig. 1.19   Portrait of Isaac Newton by Sir Godfrey Kneller

Fig. 1.20   Portraits of Jacob (*left photo*) and Johann Bernoulli by unknown authors

Fig. 1.21 Portrait of Daniel Bernoulli by Johann Jacob Haid

mechanics of rigid bodies (Newton had confined his models to geometric proofs), as well as a rigorous explanation of conservation of mass, and the introduction of strain.

It remained for Joseph-Louis Lagrange (1736–1813) (Fig. 1.23) to place the mechanics of rigid bodies on an essentially complete foundation that was both correct and clear. Using what is now termed variational calculus, Lagrange opened the door at the end of the eighteenth century to the next great surge in the science of mechanics: the development of general three-dimensional theories for deformable bodies. For his accomplishments, Lagrange is entombed in France's most honored crypt, beneath the Pantheon in Paris.

In 1808, Pierre-Simon Laplace (1749–1827) (Fig. 1.24), one of the greatest mechanicians of the era, invited the German physicist and musician Ernst Chladni (1756–1827) (Fig. 1.25) to demonstrate a series of experiments on plates before the Paris Academy of Sciences. With great precision, Chladni demonstrated scientifically through a series of experiments wherein he strummed with a violin bow on glass plates covered with sand particles that the lines that formed on the surfaces of

Fig. 1.22 Portrait of Leonhard Euler by Johann Georg Brucker

Fig. 1.23 Portrait of Joseph-Louis Lagrange by Giuseppe Lodovico

Fig. 1.24 Posthumous portrait of Pierre-Simon Laplace by Madame Feytaud (1842)

the plates were repeatable but distinct for differing boundary conditions applied to the plates as shown in Fig. 1.26.

The Emperor Napoleon attended these demonstrations, and he was so impressed that in 1809 he set a prize of 20,000 French francs to the first person who could develop a model capable of predicting the experimental results obtained by Chladni. The competition was to last 3 years, and at the end only a single entry had been submitted. That entry was written by a woman—Sophie Germain (1776–1831) (Fig. 1.27). The revelation that no man had submitted an entry was made even more profound by the fact that women were not allowed to study in higher education in France at that time. Unfortunately, Ms. Germain's solution was erroneous, but the awards committee recommended that she continue her work on the subject. Under the tutelage of Joseph Lagrange (who died before the problem

Fig. 1.25 Portrait of Ernst
Chladni

Fig. 1.26 Photo of Chladni
plate experiment

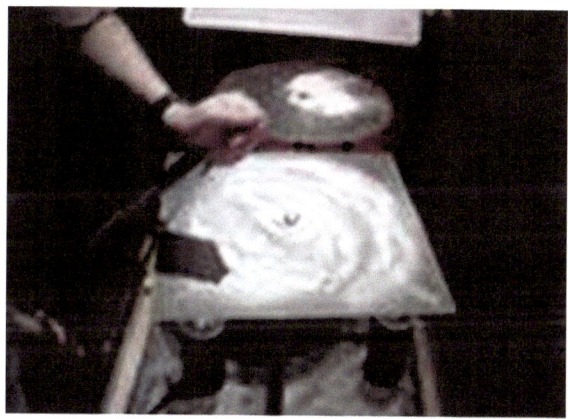

was completely solved) Ms. Germain was eventually awarded the prize in 1816, thus becoming the first woman to win a prize from the Paris Academy of Sciences (Bucciarelli and Dworsky 1980).

Although her assumptions were flawed, her solution for the plate problem was the first (essentially correct) multi-dimensional model ever reported for deformable bodies, and it paved the way for a flurry of profound developments over the succeeding decade. First, working independently, Siméon-Denis Poisson (1781–1840) (Fig. 1.28) and Claude-Louis Navier (1785–1836) (Fig. 1.29)

Fig. 1.27   Photo of bust of Sophie Germain on Rue Sophie Germain, Paris

Fig. 1.28   Portrait of Siméon-Denis Poisson

Fig. 1.29 Bust of Claude-Louis Navier

developed the modern three-dimensional theory of fluids. Shortly thereafter, Joseph Fourier (1768–1830) (Fig. 1.30), who had perhaps prophetically accompanied Napoleon to the searing heat of Egypt, developed the modern theory of heat. Finally, Navier and Augustin-Louis Cauchy (1789–1857) (Fig. 1.31) independently developed the theory of elastic solids, later further elucidated by Gabriel Lamé (1795–1870) (Fig. 1.32). While much of this work was initiated from a molecular base, they all eventually adopted a simpler framework utilized by Sophie Germain and perfected by Augustin-Louis Cauchy that assumes that the body of interest is everywhere continuous, an assumption that is today called "continuum mechanics." This approach, as well as the careful development of the modern interpretation of calculus (including the fundamental theorem of calculus) by Cauchy, laid the groundwork for the modern theories of deformable bodies.

In the midst of these developments, Baron Cauchy introduced a definition for mechanical stress that has stood the test of time, becoming the single most important concept required for the purpose of predicting failure of solids due to yielding and/or fracture (Cauchy 1822). Karl Culmann (1821–1881) would later show that the transformation of stress from one coordinate system to another could be represented graphically by a circle (Culmann 1866), and this method would be explored in

Fig. 1.30 Sketch of Joseph
Fourier by an unknown artist
c. 1820

Fig. 1.31 Photo from
Smithsonian Institution
Libraries of Augustin-Louis
Cauchy taken circa 1856 by
E.H. Reutlinger

great detail by Christian Otto Mohr (1835–1918) (Fig. 1.33), from whence we have
the graphical method for stress transformations termed Mohr's circle.

Thus, it can be seen that while no single person can claim credit for the modern
theory of deformable bodies per se, there is credit enough to go around for

Fig. 1.32 Photograph
of Gabriel Lamé

Fig. 1.33 Painting
of Christian Otto Mohr
by Osmar Schindler

Lagrange, Germain, Fourier, Navier, Cauchy, and Mohr. Further credit must be
accorded to those who are remembered for their contributions to the development of
elastic material properties: Hooke, Thomas Young (1773–1829) (Fig. 1.34), Lamé,
and Poisson.

By 1822 the problem of predicting the mechanical response of a deformable body, whether fluid or solid, had been reduced to solving a mathematical problem of predicting stresses and deformations within the body. Unfortunately, mathematical solutions for problems of this type were a hurdle that was not surmounted for more than a century. Early attempts at solutions focused on objects of specific shape, as pioneered by Jean-Claude Barré de St. Venant (1797–1886). There grew a field of applied mathematics called "elasticity theory," and this field flowered until well into the latter half of the twentieth century.

A supplementary part of the model is necessary in order to develop design methodologies capable of creating structural components that will not fail due to excessive fracture. This part of the model requires the development of failure theories. Pioneered by Charles Augustin de Coulomb (1736–1806), failure models capable of predicting yielding were developed by Henri Edouard Tresca (1814–1885), Richard von Mises (1883–1953), and many others for various materials. Models capable of accurately predicting failure due to fracture in solids were first proposed by A.A. Griffith (1920) and were dramatically improved toward the end of the twentieth century to the point that it is now possible to accurately predict fracture in many types of solid media.

Finally, the development of the high speed computer in the latter portion of the twentieth century paved the way for a solution method termed in 1960 by Ray Clough (1920) the "finite element method." This computational method for calculating the mechanical response of solid and fluid objects (as well as many other physical phenomena) has now been perfected to the point that the predictive models developed in the early nineteenth (see Fig. 1.35) can be said to have reached an advanced state of maturity.

Fig. 1.35 Chronology of French solid mechanists of the nineteenth century

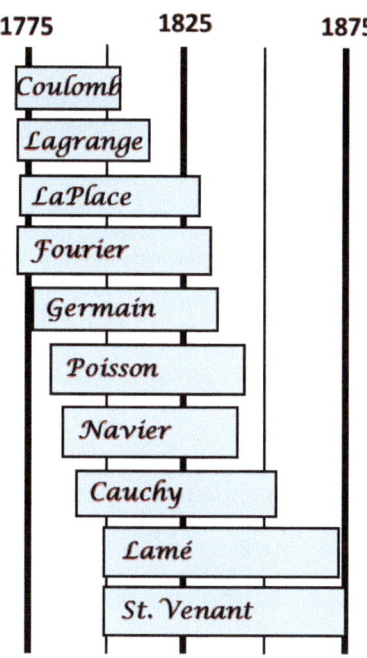

1.2 Assignments

PROBLEM 1.1

GIVEN: This chapter has discussed several famous persons who contributed to the history of mechanics.

REQUIRED: Pick any one of these persons, write a discourse describing their life and their most important accomplishment and include an image of this person.

References

Bucciarelli L, Dworsky N (1980) Sophie germain: an essay in the history of the theory of elasticity (studies in the history of modern science). Springer, New York

Cauchy A (1822) Lecture presented to the French Academy of Sciences. Unpublished, Paris

Copernicus N (1543) De revolutionibus orbium coelestium. Johann Petreius, Nuremberg

Culmann K (1866) Die graphische Statik. Verlag von Meyer & Zeller, Zurich

Descartes R (1954) La géométrie. Dover, New York

Einstein A, Calaprice A (1996) The new quotable Einstein. Princeton University, Princeton

Euler L (1744) Method inveniendi lineas curvas. Opera Omnia, St. Petersburg

Galilei G (1638) Dialogues concerning two new sciences. University of Toronto Library, Toronto

Jardine L (2005) The curious life of Robert Hooke: the man who measured London. Harper Perennial, New York

King R (2001) Brunelleschi's dome: how a renaissance genius reinvented architecture. Penguin, London

Netz R, Noel W (2009) The archimedes codex: how a medieval prayer book is revealing the true genius of antiquity's greatest scientist. Da Capo, Filadelfia

Newton I (1686) The principia: mathematical principles of natural philosophy. Prometheus, New York

Reti L (1990) Unknown Leonardo. Abradale, New York

Santayana G (1905) Reason in common sense: the life of reason Volume 1. Dover, New York

Teresi D (2003) Lost discoveries: the ancient roots of modern science—from the babylonians to the Maya. Simon & Schuster, New York

Selected Reading

Aczel A (2006) Descartes' secret notebook. Broadway, New York

Aczel A (2002) The riddle of the compass. Mariner, Wilmington

Adler K (2002) The measure of all things: seven-year odyssey and hidden error that transformed the world. Free Press, New York

Aughton P (2005) The transit of venus. Orion, London

Beckmann P (1971) A history of pi. Golem Press, New York

Bergreen L (2003) Over the edge of the world: magellan's terrifying circumnavigation of the globe. William Morrow, New York

Bolles E (1997) Galileo's commandment an anthology of great science writing. Freeman, New York

Boyer C (1959) The history of calculus and its conceptual development. Dover, New York

Bryson B (2003) A short history of nearly everything. Broadway Books, New York

Burleigh N (2007) Mirage: Napoleon's scientists and the unveiling of Egypt. Harper, New York

Cahill T (2008) Mysteries of the middle ages and the beginning of the modern world. Anchor, New York

Capra F (2007) The science of leonardo. Doubleday, New York

Connor J (2004) Kepler's witch. HarperCollins, New York

Cutler A (2003) The seashell on the mountaintop. Heinemann, London

Daintith J, Gjertsen D, Ed. (1999) A dictionary of scientists. Oxford Press, Oxford

De Camp L (1993) The ancient engineers. Barnes & Noble, New York

Derbyshire J (2003) Prime obsession bernhard riemann and the greatest unsolved problem in mathematics. Joseph Henry, Washington

Devlin K (2011) The man of numbers Fibonacci's arithmetic revolution. Walker, New York

Diamond J (1992) The third chimpanzee. HarperCollins, New York

Diamond J (2005) Guns, germs and steel a short history of everybody for the last 13,000 years. Norton, New York

Diamond J (2004) Collapse: how societies choose to fail or succeed. Viking, New York

Drake S (2003) Galileo at work his scientific biography. Dover, New York

Dugas R (1988) A history of mechanics. Dover, New York

Duncan D (1998) The calendar. Fourth Estate, London

Dunham W (1999) Euler the master of us all. Mathematical Association of America, Washington, DC

Ferguson K (2002) Tycho and Kepler The strange partnership that revolutionised astronomy. Walker, New York

Ferris T (1991) The world treasury of physics, astronomy, and mathematics. Little, Brown & Co., Boston

Feynman R (1988) What do you care what other people think? Norton, New York

Feynman R (1997) Surely you're joking Mr. Feynman. Norton, New York

Galison P (2003) Einstein's clocks, poincare's maps. Norton, New York

Gamow G (1988) The great physicists from Galileo to Einstein. Dover, New York

Gilder J, Gilder AL (2005) Heavenly intrigue. Anchor, New York

Gillispie C (1997) Pierre-Simon Laplace 1749–1827: a life in exact science. Princeton University Press, Princeton

Gingerich O (2004) The book nobody read chasing the revolution of nicolaus copernicus. Walker, New York

Gleeson J (1999) The arcanum. Little, Brown & Co., Boston

Gleick J (2004) Isaac Newton. Vintage, New York

Goldstone L, Goldstone N (2005) The friar and the cipher: roger bacon and the unsolved mystery of the most unusual manuscript in the world. Doubleday, London

Grabiner J (2011) The origins of Cauchy's rigorous calculus. Dover, New York

Gribbin J, Hook A (2003) The scientists. Random House

Hawking S (1998) A brief history of time. Bantam Doubleday Dell, New York

Hawking S (2005) God created the integers. Running, Philadelphia

Hellman H (1998) Great feuds in science. Wiley, New York

Herman A (2001) How the Scots invented the modern world. Crown, New York

Hoffman P (2003) Wings of madness Alberto Santos-Dumont and the invention of flight. Hyperion, New York

Holmes R (2009) The age of wonder: how the romantic generation discovered the beauty and terror of science. Pantheon, London

Hough R (1995) Captain james cook: A biography. Norton, New York

Isaacson W (2008) Einstein his life and universe. Simon & Schuster, New York

King R (2006) The judgment of Paris. Walker, New York

Landels J (2000) Engineering in the ancient world. University of California Press, Berkeley

Linklater A (2003) Measuring America. Penguin, New York

Mahon B (2003) The man who changed everything the life of James clerk Maxwell. Wiley, Hoboken

Manchester W (1992) A world lit only by fire: the medieval mind and the renaissance. Little, Brown & Co., Boston

Metz B (2007) Temples, tombs and hieroglyphs. William Morrow, New York

Moorehead A (2000) The white nile. Harper, New York

Moorehead C (1994) Lost and found the 9,000 treasures of troy. Weidenfeld & Nicolson, London

Parsons W (1939) Engineers and engineering in the renaissance. Williams and Wilkins, Baltimore

Pumfrey S (1999) Latitude and the magnetic earth. Icon Books, Cambridge

Reston J (2005) Galileo: a life. Beard, Santa Ana, CA

Robinson A (2005) The last man who knew everything. Pi Press, New York

Sagan C (1980) Cosmos. Random House, New York

Sagan C (1994) Pale blue dot. Random House, New York

Shlain L (1991) Art and physics: parallel visions in space, time and light. William Morrow, New York

Shrady N (2003) Tilt: A skewed history of the tower of pisa. Simon & Schuster, New York

Simmons J (2009) The Scientific 100. Fall River Press, New York

Sobel D (1999) Galileo's daughter. Walker, New York

Sobel D (2005) Longitude: the true story of a lone genius who solved the greatest scientific problem of his time. Walker, New York

Stein S (1999) Archimedes what did he do besides cry eureka? Mathematical Association of America, Washington, DC

Timoshenko S (1953) History of the strength of materials. McGraw-Hill, New York

Timoshenko S (1968) As I remember. Van Nostrand, Princeton

Toddhunter I, Pearson K (1893) A history of the theory of elasticity. Cambridge University Press, Cambridge

Truesdell C (1968) Essays in the history of mechanics. Springer, New York
Tyldesley J (2006) EGYPT how a lost civilization was rediscovered. University of California Press, Berkeley
Uglow J (2002) The lunar men. Farrar, Straus & Giroux, New York
Walker P (2002) The feud that sparked the renaissance. William Morrow, New York
Winchester S (2001) The map that changed the world. Harper, New York
Zebrowski E Jr (1999) A history of the circle. Rutgers University Press, New Burnswick

Chapter 2
Mechanics of Materials

2.1 Introduction

This text is concerned with the mechanics of solids. Mechanics is the study of the motion of bodies. A solid is defined to be an object that retains its shape when it is unloaded and unconfined. A body that deforms without being loaded is called a fluid. Necessarily, fluids do not undergo fracture, as they are for all intents and purposes fractured by the absence (or paucity) of molecular bonds. Solids, on the other hand, can and do undergo fracture when subjected to loading conditions sufficient to induce and/or propagate cracks. Sometimes these cracks can accumulate and/or grow in such a way as to cause the object to fail in the sense that it no longer is capable of performing its intended function. ***This then, is a major objective of this text—to develop models that can be used to design solid objects that are capable of withstanding fracture (as well as other modes of failure) when subjected to mechanical loading.***

In the previous chapter, we learned that people have been studying mechanics for at least several millennia, both as it pertains to the motions of the heavenly bodies and as it pertains to building construction. A body deforms when the distance between any two points in the body changes. Commencing with Galileo Galilei's text *Two New Sciences,* significant effort has been devoted by scientists and engineers to the development of models for predicting the response of deformable bodies (Timoshenko 1953).

Solid objects utilized for structural purposes can fail due to a variety of causes such as excessive deformations, buckling, excessive cost, and fracture. While all of these need to be considered in the design process, the failure mechanism that will be considered in detail in this text is fracture. Failure of structures due to fracture is by no means a new subject. Indeed, of the seven ancient wonders of the world, only one of these wonders remains more or less intact today—the great pyramids. The others failed long ago, usually due to fracture.

D.H. Allen, *Introduction to the Mechanics of Deformable Solids: Bars and Beams,*
DOI 10.1007/978-1-4614-4003-1_2, © Springer Science+Business Media New York 2013

Fig. 2.1 Photo of Stonehenge on the Salisbury Plain in Southern England

Fig. 2.2 Street scene in Pompeii

Early engineers undoubtedly had some rudimentary understanding of the cause of fracture, as evidenced by the beam–arch–column structures shown in Figs. 2.1, 2.2, and 2.3.

Stonehenge, on the Salisbury plain in southern England, is apparently a very old site, dating back possibly as far as 8,000 BC. Carbon dating of wooden fragments

Fig. 2.3 Photo of the Canopus at Hadrian's Villa

found inside the dirt berm suggests that the structure may have originally been wooden. It is likely that early engineers realized that in order to build structures that would stand the test of time, a more durable material was needed—stone. And indeed, almost all structures remaining today from antiquity are stone. Unfortunately, as demonstrated by the large horizontal stones in Fig. 2.1, the low tensile strength of stone limits its use as a beam of substantial length.

Excavations at Pompeii have revealed roofless structures virtually everywhere within the city destroyed by the eruption of Vesuvius in 79 AD, as shown in Fig. 2.2. As stone construction was extremely expensive even then, most structures used wooden beams for roof structures in Pompeii.

The Romans invented the arch in order to provide larger spans for stone structures, and this invention allowed the Romans to create many of the most famous structures still standing today from that time period. A telling example is the portico from the Canopus at Hadrian's villa in Tivoli, built in the second century AD (Fig. 2.3). The portico has both flat and curved stone members between the arches, and the discerning reader will recognize that the span between the arches is slightly larger than that between the beams, attesting to the fact that arches can span larger dimensions than beams made of stone because they carry loads strictly in compression, whereas beams necessarily undergo tensile loading on one side or the other, a circumstance that precludes the use of stone for large spans.

The Romans expanded this understanding of compression to build perhaps their most amazing structure—a dome—in the second century, again during the reign of Hadrian. The Pantheon, the last completely intact Roman structure in Rome, stands today as a monument to the ingenuity of the Romans, as shown in Fig. 2.4.

It is not known today exactly who undertook the reconstruction of the old temple built by Augustus Caesar's close friend Marcus Agrippa, but what is certain is that it

Fig. 2.4 Photos of the Pantheon, exterior photo on *left*, and interior photo on *right* showing the oculus

Fig. 2.5 Photo of the Pont du Gard (note people standing on the lower deck)

was revolutionary. The dome is made of concrete, a technology that was lost after the fall of the Roman Empire in the fifth century until the nineteenth century, when the French reinvented concrete technology. A careful study of this structure will lend credence to the enormous impact that the Romans had on western civilization.

Despite their proven ability to construct both massive and impressive structures, ancient engineers did not possess rigorous design methodologies. Theirs was an experimental and necessarily expensive discipline. For example, it is known that the Pont du Gard, built in the first century AD (Fig. 2.5), was constructed at a cost that would have bankrupted a small nation today. It is noteworthy that this massive aqueduct in the south of France still stands today, so that the cost may not sound so impressive if amortized over two millennia.

2.2   Modern Models

The models we use today to design structures are robust, in the sense that essentially all of the controllable inputs can be manipulated analytically (meaning—*without actually having to build the structure!*) in order to produce an acceptable design. More importantly, today's models have been shown repeatedly through careful experimentation to be accurate. All of the models that will be developed in this text are built on three important but distinct types of variables: *independent variables, input variables, and output variables.*

Independent variables are composed of time and spatial coordinates. In this course it will be assumed that structural response is time independent, so that the first of these independent variables will not appear in our models. Spatial coordinates will normally appear via an assigned coordinate system, such as Cartesian coordinates (x, y, z), after the French mathematician Descartes. It should be pointed out that choosing both the origin and the orientation of a coordinate system is completely arbitrary and is therefore at the discretion of the modeler. However, in this course the following convention will be employed throughout:

1. *The coordinate origin will always be placed at the left end of the long axis of the object.*
2. *The coordinate system will always be right handed.*
3. *The y coordinate direction will (almost) always be placed normal to the x-axis and in the plane of the page.*

Input variables in all mechanics problems are of three distinct types: *loads, geometry, and material properties.* These compose the complete set of necessary information that must be known before a structural analysis can be carried out. For example, the shape of the structure must be known a priori, before the analysis can be performed, and this is termed the geometry of the structure.

Output variables are the set of items that result from the model development. They generally consist of kinetic (such as stress, to be defined below) and kinematic quantities (such as displacements). Once a cogent model has been constructed, the output variables will appear as explicit functions of the independent variables and the input variables. Thus, for example, in a beam, the displacement field will be modeled as a function of the input loads, the material properties of the beam, and the shape of the beam.

As a consequence of the continuous nature of structural components, the resulting models in this course will employ differential calculus, so that the models will be at least in part in the form of differential equations (Malvern 1969; Glover and Jones 1992). Thus, it will be necessary to solve these equations for specific sets of loads, geometry, and material properties so as to describe them via simpler algebraic equations. This type of robust model can then be inverted in such a way that the input loads, geometry, and material properties can be *designed* so as to create a structure that will satisfy any and all design constraints. The subject of structural design will be addressed in Chap. 7.

In this chapter, we develop (and in some cases review) the fundamental mechanics that are required in order to develop models capable of predicting the response of solids to mechanical loadings. These fundamentals fall into three general classes: kinetics, kinematics, and constitution. Insofar as they relate to the current subject matter, these are discussed in some detail below.

2.2.1 Kinetics

Kinetics is the study of mechanical loads acting on objects. There are two fundamentally different ways that mechanical loadings can be imparted to bodies: via body forces (expressed in force per unit volume) or via surface tractions (expressed in force per unit area). For example, when a person sits still in a chair, the mass of the person results in a force per unit volume that acts on every mass point (in the interior of the person) in the direction of the center of mass of the Earth. This force, F, is called a gravitational force, and it is directly proportional to the mass, m_1, of the person, the mass, m_2, of the earth, and inversely proportional to the square of the distance between the mass point and the center of gravity of the earth, r. The law describing **body forces** was first espoused rigorously by Isaac Newton in his book *The Principia*, and for that reason it is called Newton's gravitational law, given by

$$F = G \frac{m_1 m_2}{r^2} \tag{2.1}$$

where F is the magnitude of the force, G is the gravitational constant, and the direction of the force is through the straight-line connecting the mass point in question to the center of mass of the earth.

Unlike body forces, ***surface tractions*** (expressed in units of force per unit area and to be defined below) act on the surface of objects when they come in contact with other objects. From Newton's third law, we know that two objects that are in contact with one another exert equal and opposite forces on one another. It has also been proven (by Augustin Cauchy) that two objects in contact with one another exert equal and opposite surface tractions on one another, meaning that not only are the forces identical, but the distribution of those forces is also identical between the two bodies in contact.

When a person sits in a chair, the surface tractions act on the part of the person's body that comes in contact with the chair, and those acting on the chair are of identical magnitude and opposite sign to those acting on the person. So why doesn't the person in the chair fly upwards due to the force being applied to him or her by the chair? The answer is that the resultant of the surface tractions upwards (caused by the chair pushing on the person) is exactly equilibrated by the resultant of the body force downwards (equal to the weight of the person), and this is described by Newton's first law (sometimes called conservation of momentum):

$$\sum \vec{F} = 0, \quad \sum \vec{M} = 0 \tag{2.2}$$

If the body is in motion, then the right hand side of the above two equations is not zero, but is expressed as the rate of change of the momentum, called Newton's second law. In this course, we will consider only objects that are in equilibrium, so that (2.2) is sufficient to guarantee that momentum is conserved. The above laws describe the kinetics of all bodies in the universe that are at rest.

There are several other conservation laws that have been espoused over the last three centuries. These include: (1) conservation of mass, (2) conservation of energy, (3) conservation of charge, and (4) the entropy production inequality. These laws are for most practical circumstances demonstrated to be true everywhere on our planet. Therefore, the discerning reader may ask why these laws are not utilized in the current text. The answer is that they are not needed because in all of the circumstances that will be considered herein they are all trivially satisfied, meaning that they provide no additional useful information, while still being true. Therefore, they will not be considered further herein.

The concept of load intensity has long been recognized as a means of estimating the load carrying ability of a solid. A rough approximation of load intensity can be constructed by dividing the total load acting on a surface by the area of the surface. For example, the load intensity of the great pyramid of Cheops is equal to the total force caused by the mass of the pyramid in earth's gravitational field divided by the footprint of the base of the pyramid. This force is equal to the mass of the pyramid multiplied by Earth's gravitational constant, g. A further example of this concept can be seen in Fig. 2.6, wherein a truncated pyramid of weight (force), W, is shown resting on a plane both upright and inverted.

It is clear from Fig. 2.6 that the average pressure on the base of the inverted pyramid, p_I, is greater than the pressure on the upright pyramid, p_U, because the weight of the pyramid is the same whether it is upright or inverted, and $b > a$. Thus, if the pyramid is resting on a soft base, such as sand (in the Egyptian desert!), it is much more likely to cause failure of the base material if it is inverted than if it is upright. Thus, it is clear *that load **intensity** is more important for predicting failure than is load itself.*

A terminology has been developed with respect to this load intensity and it is called traction. The traction acting on a surface, as shown in Fig. 2.7, is defined as follows.

$$\vec{t}(\vec{n}) \equiv \lim_{\Delta A \to 0} \frac{\overrightarrow{\Delta F}}{\Delta A} \tag{2.3}$$

Fig. 2.6 Average pressure
on the base of a truncated
pyramid

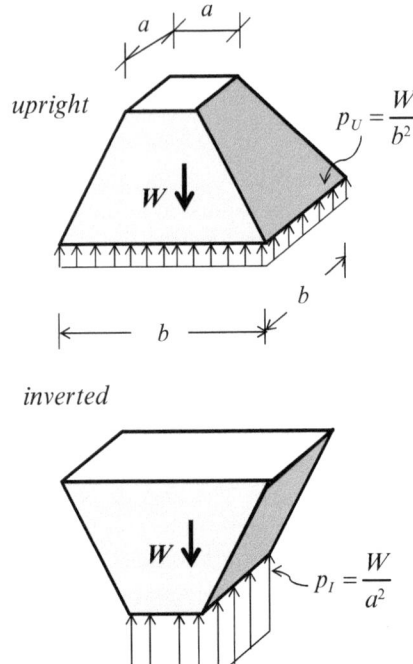

Fig. 2.7 Traction, $\vec{t}$, on
plane P, with unit outer
normal vector, $\vec{n}$, in body B

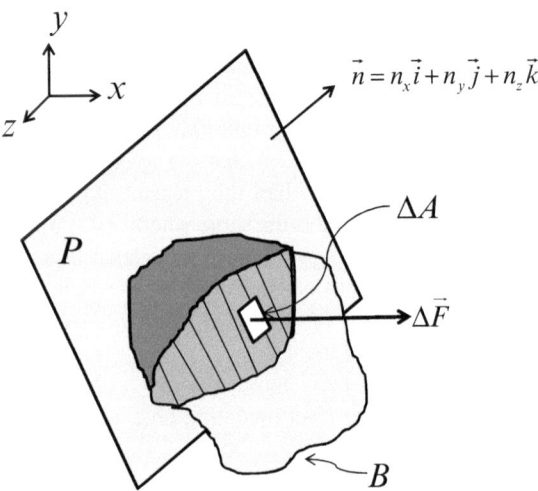

Where, as shown in the figure, $\vec{n}$ is the unit outer normal vector to the plane, ΔA is the area of the plane, and $\Delta \vec{F}$ is the force acting on the plane. Note that $\vec{t}(\vec{n})$ may be written in its component form as follows:

$$\vec{t}(\vec{n}) = t_x\vec{i} + t_y\vec{j} + t_z\vec{k} \tag{2.4}$$

Fig. 2.8 Physics of different
types of mechanical boundary
conditions

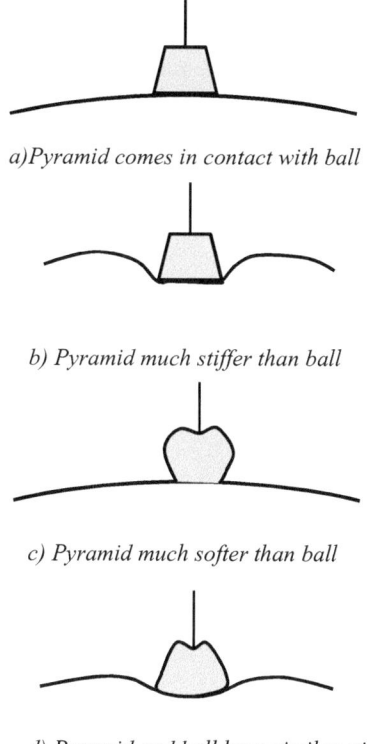

a)Pyramid comes in contact with ball

b) Pyramid much stiffer than ball

c) Pyramid much softer than ball

d) Pyramid and ball have similar stiffnesses

where $\vec{i}, \vec{j}$, and $\vec{k}$ are unit base vectors in the x, y, and z coordinate directions, respectively.

It can be seen from the above definition that the traction vector is described in units of force per unit area. Thus, if the traction vector is normal to the surface of interest, it is equivalent to a pressure, but it can also have components parallel to the surface of interest. Furthermore, the traction vector may act on any plane (defined by the unit outer normal vector, $\vec{n}$) and at any point inside or on the surface of an object.

2.2.2 Boundary Conditions

Suppose that the plane of interest is chosen to coincide with the boundary of the object in question. Due to the physics of the problem, it is necessary to develop a mathematical model describing what is occurring at the boundary, termed the so-called boundary conditions. In order to describe the physics, consider a block that is lowered slowly onto a curved surface until the two objects touch at a single point, as shown in Fig. 2.8a. As the block continues to be lowered, there are three possible results (assuming that the block is held in place over the curved surface):

CASE 1: The surface conforms to the shape of the block (Fig. 2.8b)
The block is obviously much stiffer than the surface, such as the case of a steel block resting on a foam rubber surface. In this case, the foam rubber surface conforms to the shape of the block, resulting in what is termed traction boundary conditions applied to the boundary of the block where it is in contact with the surface.

Traction boundary conditions: $\vec{t}(\vec{n}) = $ *known on the boundary of the block*

CASE 2: The block conforms to the shape of the surface (Fig. 2.8c)
The surface is clearly much stiffer than the block, such as the case of a foam rubber block resting on a steel surface. In this case, the block conforms to the shape of the surface, resulting in what is termed displacement boundary conditions applied to the block where it is in contact with the surface.

Displacement boundary conditions: $\vec{u} = $ *known on the boundary of the block*

CASE 3: Both the block and the surface deform significantly (Fig. 2.8d)
Both the block and the surface are made of the same (or similar) materials. This case, called a structural interaction problem, requires that both objects be analyzed simultaneously. Where the two come in contact with one another, the boundary conditions are not known. All that is known is that both the shapes and tractions must match, called matching conditions.

We will not consider this last possibility in this course, as it is an advanced subject that is beyond the scope of this text.

The reader will perhaps understand the above discussion best if the example of his/her own body is considered. If a person sits on pavement, his/her body will conform to the shape of the pavement where there is direct contact, and this part of the person's exterior is subjected to displacement boundary conditions. Note that sitting on the pavement is generally discomforting, and this is due to locally large stresses within the person near the points of contact. Note also that the part of the person's exterior that is not in contact with the pavement is subjected to air pressure. Since air is relatively compliant compared to the person, this part of the person's exterior is subjected to traction boundary conditions. Conversely, if a person is placed on a relatively compliant object, such as a water bed, then the bed will conform to the shape of the person's exterior, and the person is therefore subjected to traction boundary conditions on the portion of their exterior that is in contact with the bed. This, of course, explains in large measure why beds are more comfortable than pavement—stress concentrations are largely mitigated by traction type boundary conditions.

As we will see later in the text, both traction and displacement boundary conditions occur often in the analysis of solids. However, due to the physics described above, only one type of boundary condition is possible in each coordinate direction at a point on the surface of an object.

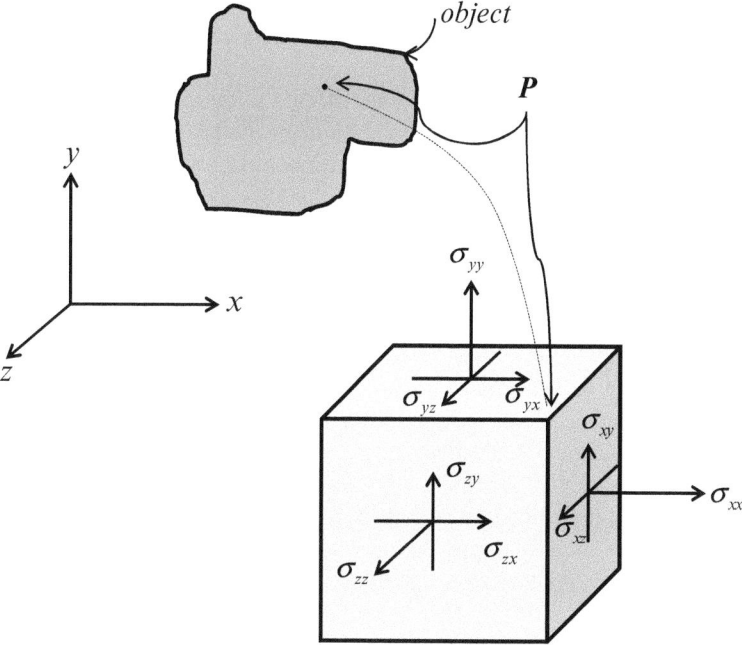

Fig. 2.9 Components of the stress tensor at a point P in an object

2.2.3   The Stress Tensor

Now let us examine the interior of the object of interest. Suppose that three planes are chosen so as to pass through an arbitrary point P in the interior of the object and perpendicular to the three coordinate axes. Then the traction vector on these three planes will take on a particular significance, as defined by Augustin Cauchy. In this case, the nine resulting components are called the Cauchy stress tensor, as shown in Fig. 2.9, and are given by

$$\vec{t}(\vec{i}) \equiv \sigma_{xx}\vec{i} + \sigma_{xy}\vec{j} + \sigma_{xz}\vec{k}$$
$$\vec{t}(\vec{j}) \equiv \sigma_{yx}\vec{i} + \sigma_{yy}\vec{j} + \sigma_{yz}\vec{k}$$
$$\vec{t}(\vec{k}) \equiv \sigma_{zx}\vec{i} + \sigma_{zy}\vec{j} + \sigma_{zz}\vec{k} \qquad (2.5)$$

The sign convention for subscripts on the nine components of the stress tensor described above can be seen to be as follows: *the first subscript is associated with the unit normal vector for that plane, and the second subscript is the direction that the stress component is oriented.* Note also that if the subscripts are the same for any component of stress, that component is perpendicular to the plane of interest and is therefore termed a *normal stress*. If the subscripts are not the same for any component of stress, that component is parallel to the plane of interest and is therefore termed a *shear stress*. Finally, the stress is called a tensor because the orientation of the components of the stress transform in a very complicated way (unlike the components of a vector), as we will see in Chap. 6.

Equation (2.5) is a very important definition in the history of mechanics. Monsieur Cauchy is responsible for many important developments—this is one of his very best. As we will see, the ability to predict the components of stress at all points in a body of arbitrary shape is at the very heart of structural design. Cauchy went a step further with the above definition of stress by using Newton's first law (summing forces) to prove *that the nine components of stress at a point in an object are sufficient to uniquely define the state of loading at that point in the object*. Therefore, it has been established that the stress tensor is the key kinetic quantity necessary to determine the state of loading at every point in a body. While the significance of this was not immediately apparent to the engineering community when Cauchy first reported it in 1822, it was clear by the middle of the nineteenth century that if the stress tensor could be predicted at every point in an object, it could be utilized as a means of predicting whether or not the body would be capable of withstanding the loads applied to it. Thus, it may be said that *stress is the most important concept underlying all of modern structural and solid mechanics*.

The discerning reader may ask the question—how can there be stress at the molecular or atomistic scale? Of course, the answer is that there really is no such thing as a continuum, from whence the concept of stress emanates. In fact, at the scale that we normally employ it stress is nothing more than an ensemble average of molecular and atomistic forces per unit area. As such, it does not exist in reality. And furthermore, it cannot be measured. It can only be inferred by measuring displacements and employing constitutive equations, to be described below. Nevertheless, this ingenious concept has been shown through experimental observation to be a powerful means of predicting failure due to yielding, buckling, creep, excessive deformations, and fracture. Interestingly, Sophie Germain and Augustin Cauchy appear to be among the first to develop their models for media by starting from the assumption that the media can be idealized as continua. Prior to their attempts, such as the work of Navier, models proceeded from the molecular scale. For cases where the object of interest is large compared to the molecular scale, these latter approaches have given way to the continuum approach employed by Germain and Cauchy.

By summing moments at an arbitrary point in an object it can also be shown that the stress tensor is symmetric, i.e.,

$$\sigma_{xy} = \sigma_{yx}, \ \sigma_{xz} = \sigma_{zx}, \ \sigma_{yz} = \sigma_{zy} \tag{2.6}$$

Thus, there are only six unique components of stress at a given point in an object.

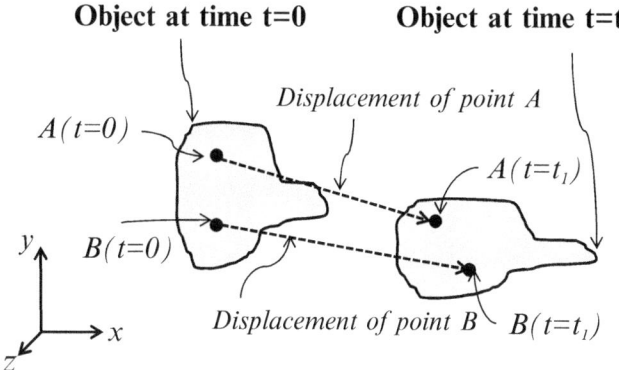

Fig. 2.10 Depiction of two points in a deformable body

2.2.4 Kinematics

Kinematics is the study of the motions of objects without reference to the forces involved. Three-dimensional objects sometimes move in such a way that the object in question may be considered for practical circumstances to be rigid. The term rigid implies that the object does not deform, so that all materials points in the object retain their relative distances from one another throughout the motion of the object. This is clearly an approximation to reality, and this approximation is not sufficient for determining whether the body will fail due to deformations. The current course is focused on deformable body motions, so that the case of rigid body motions will not be considered herein.

When a body deforms, each material point may undergo a distinct path of motion in time, as demonstrated in Fig. 2.10 for two points in a typical deformable body.

It is clear from Fig. 2.10 that the displacement vector, $\vec{u}$, is a function of coordinate location in the body, that is, $\vec{u} = \vec{u}(x, y, z, t)$, where the components of the displacement vector are given by

$$\vec{u}(x, y, z, t) = u(x, y, z, t)\vec{i} + v(x, y, z, t)\vec{j} + w(x, y, z, t)\vec{k} \qquad (2.7)$$

Because the displacement vector is a function of position, the spatial gradient of the displacement will not necessarily be zero in a deformable body. Therefore, suppose that a new variable is introduced, called the strain tensor, with the following components:

$$\varepsilon_{xx} \equiv \frac{\partial u}{\partial x}, \varepsilon_{yy} \equiv \frac{\partial v}{\partial y}, \varepsilon_{zz} \equiv \frac{\partial w}{\partial z}$$

$$\varepsilon_{xy} \equiv \frac{1}{2}\left(\frac{\partial u}{\partial y} + \frac{\partial v}{\partial x}\right), \quad \varepsilon_{xz} \equiv \frac{1}{2}\left(\frac{\partial u}{\partial z} + \frac{\partial w}{\partial x}\right), \quad \varepsilon_{yz} \equiv \frac{1}{2}\left(\frac{\partial v}{\partial z} + \frac{\partial w}{\partial y}\right) \qquad (2.8)$$

where we have employed the symbol $\equiv$ to mean "is defined to be."

Fig. 2.11 A uniaxial test

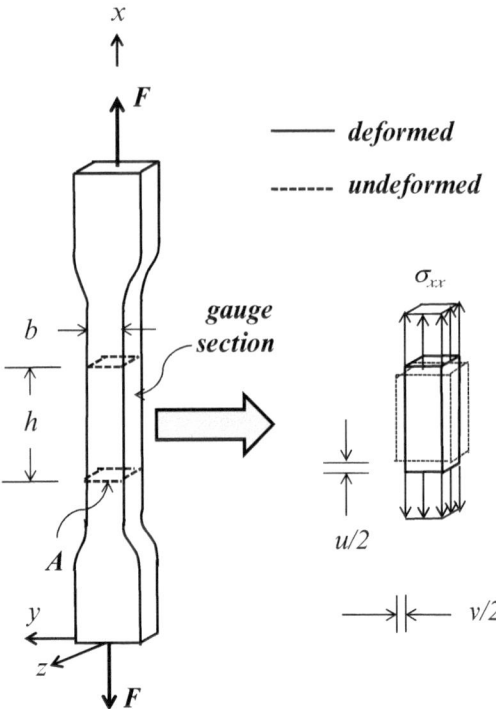

As has been discussed above, when a body translates as a rigid body, the displacements at all points in the body are equivalent, so that all of the components of the strain tensor defined in (2.8) will be identically zero. Thus, the definition of strain given above can be viewed as a means of filtering out rigid body translations. The importance of this revelation will become important in the next section.

2.2.5   Material Behavior

In order to complete the development of a rigorous model for predicting the mechanical response of structural components, it is necessary to develop a model describing the material behavior of the component to be modeled. Such a model requires the construction of a well-designed experiment called a constitutive test, to be defined below. Toward this end, the experiment originally proposed by Leonardo Da Vinci in Chap. 1 is useful for supplying a wealth of information that is relevant to this text. A modern version of Da Vinci's experiment consists of applying a load to the end of a prismatic bar composed of the material in question and measuring the deformation as a function of the applied load, as shown in Fig. 2.11. ***This type of test is termed a uniaxial test*** because the load is applied parallel to the long axis of the bar (Allen and Haisler 1985). Of course, it is not possible to measure load directly in the laboratory (or anywhere else, for that matter). Thus, this is accomplished indirectly by placing a device in series with the specimen called a load cell, which is nothing more than a fancy name for a spring that obeys Hooke's law. The load cell can then be used to deduce the load, and a variety of techniques (such as strain gauges, linear voltage differential transducers (LVDTs), or optical measuring devices) may be used to measure the displacement (and resulting axial strain) in the bar. This test then results in what is called an inverse problem.

It can be shown that the stresses and strains in this experiment are spatially homogeneous in the so-called gauge section, as shown in Fig. 2.11 due to something called St. Venant's principle and Newton's laws so long as the displacements are measured at least as far as the dimension b from the shoulder in the specimen. Thus, using the definitions of stress and strain given earlier in this chapter, it can be deduced that

$$\sigma_{xx} = \frac{F}{A} \tag{2.9}$$

$$\varepsilon_{xx} = \frac{u}{h} \tag{2.10}$$

and

$$\varepsilon_{yy} = \frac{v}{b} \tag{2.11}$$

where the quantities in the above equations are as shown in Fig. 2.11.

Therefore, this is a rather unique situation in which the unknowns in the problem are deducible from the loads and geometry of the specimen, and for that reason this test is termed *a constitutive test*. Thus, a plot of the axial stress versus the axial strain is easy to construct, and from this it is possible to deduce the relation between the axial stress and the axial strain.

There are other tests that may be used in this way to deduce material properties, but this test is generally the simplest (and least expensive) to perform in the laboratory for many structural materials. There are some solids for which this type of test is impractical due to the makeup of the material. Examples include certain types of soils that cannot carry significant tensile load. Concrete is another example. For materials of this type, more complicated tests are necessary. Due to their complexity, they will not be covered in this text. However, the principle of all *constitutive tests* is identical to that of a uniaxial test—to devise a means of deducing the stresses and strains inside the object directly from experimentally determined information on the boundary of the specimen.

An example of the observed relation between the stress and strain for the case of constant loading rate is shown in Fig. 2.12. Metals tested at temperatures below about 30 % of their melting temperature will typically behave in a way that is termed elasto-plastic due to the fact that they will display a relationship between stress and strain that is linear so long as the stress does not exceed a critical value called the yield point, and denoted in this text by, σ^T, which is a material constant.

Fig. 2.12 Typical uniaxial
stress–strain curve for a metal

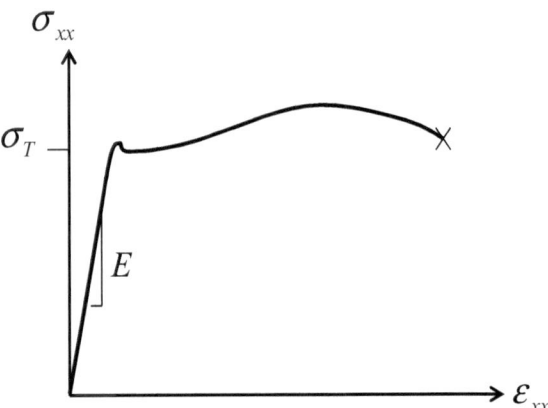

If the test is performed in compression, then the yield strength is denoted by σ^C. The yield strengths in tension and compression are often equivalent in metals. However, in other materials, such as soils and concrete, the compressive yield strength is normally much larger than it is in tension.

Up to this value, it is known that essentially all of the deformation is caused by aggregated molecular bond stretching, which is approximately linear and recoverable. Beyond this point, the behavior is not linear, and the material will undergo permanent deformation. Other responses of materials are possible. It would be nice if we could make materials behave the way we want them to, but that simply is not the case. Try as we may, they will behave as they wish. Thus, there are numerous different material models that may be required to model the stress–strain behavior, even in a simple uniaxial test. Human tissue, for example, behaves in a highly nonlinear and rate-dependent way when loaded uniaxially.

Such materials are beyond the scope of this text. Herein, we will consider only materials wherein the stress–strain behavior is linear, called Hookean, after the first person to notice this effect, Robert Hooke.

When materials behave linearly and the relation between the stress and strain is unique, the material is termed "linear elastic." In this case, the uniaxial stress–strain relation is described by

$$\varepsilon_{xx} = \frac{\sigma_{xx}}{E} \Leftrightarrow \sigma_{xx} = E\varepsilon_{xx} \qquad (2.12)$$

where the x-axis is aligned with the loading direction, as shown in Fig. 2.11. The symbol E is clearly a measurable quantity that results directly from (2.9) and (2.10) and is represented by the slope of the uniaxial stress–strain curve shown in Fig. 2.12. E is called Young's modulus, after Thomas Young. The lateral strain may also be measured in the same test, as described by (2.11), and this measurement, together with that obtained from (2.10), may be used to obtain the following material constant, called Poisson's ratio, after Siméon Denis Poisson:

$$v \equiv -\frac{\varepsilon_{yy}}{\varepsilon_{xx}} \tag{2.13}$$

A generalized three-dimensional representation of (2.12) for a generally aniso-tropic linear elastic material is given by the following:

$$
\begin{Bmatrix} \varepsilon_{xx} \\ \varepsilon_{yy} \\ \varepsilon_{zz} \\ \varepsilon_{yz} \\ \varepsilon_{xz} \\ \varepsilon_{xy} \end{Bmatrix} =
\begin{bmatrix}
C_{11} & C_{12} & C_{13} & C_{14} & C_{15} & C_{16} \\
C_{12} & C_{22} & C_{23} & C_{24} & C_{25} & C_{26} \\
C_{13} & C_{23} & C_{33} & C_{34} & C_{35} & C_{36} \\
C_{14} & C_{24} & C_{34} & C_{44} & C_{45} & C_{46} \\
C_{15} & C_{25} & C_{35} & C_{45} & C_{55} & C_{56} \\
C_{16} & C_{26} & C_{36} & C_{46} & C_{56} & C_{66}
\end{bmatrix}
\begin{Bmatrix} \sigma_{xx} \\ \sigma_{yy} \\ \sigma_{zz} \\ \sigma_{yz} \\ \sigma_{xz} \\ \sigma_{xy} \end{Bmatrix} \tag{2.14}
$$

where the C matrix can be shown to be symmetric using thermodynamic constraints and thus contains 21 unique constants. Therefore, it might be necessary to perform quite a few experiments in order to obtain all of the coefficients in the C matrix. Fortunately, there are no practical circumstances where materials are generally anisotropic. The most general case of material anisotropy commonly found is called orthotropic material behavior. Such is the case for continuous fiber composites, such as those used in the airframes of the Boeing 787 Dreamliner and the Airbus A380, as well as downhill skies, golf clubs, fishing rods, and tennis rackets, to name a few. For materials such as this, it can be shown that (2.14) simplify to the following form:

$$
\begin{Bmatrix} \varepsilon_{xx} \\ \varepsilon_{yy} \\ \varepsilon_{zz} \\ \varepsilon_{yz} \\ \varepsilon_{xz} \\ \varepsilon_{xy} \end{Bmatrix} =
\begin{bmatrix}
C_{11} & C_{12} & C_{13} & 0 & 0 & 0 \\
C_{12} & C_{22} & C_{23} & 0 & 0 & 0 \\
C_{13} & C_{23} & C_{33} & 0 & 0 & 0 \\
0 & 0 & 0 & C_{44} & 0 & 0 \\
0 & 0 & 0 & 0 & C_{55} & 0 \\
0 & 0 & 0 & 0 & 0 & C_{66}
\end{bmatrix}
\begin{Bmatrix} \sigma_{xx} \\ \sigma_{yy} \\ \sigma_{zz} \\ \sigma_{yz} \\ \sigma_{xz} \\ \sigma_{xy} \end{Bmatrix} \tag{2.15}
$$

When a material is tested, it is often observed that the test may be performed along any orientation and the results are the same. This type of response is called isotropic. Many structural materials are isotropic, and we will confine our models to isotropic materials in the current text. In this case, (2.15) simplifies to the following:

$$
\begin{Bmatrix} \varepsilon_{xx} \\ \varepsilon_{yy} \\ \varepsilon_{zz} \\ \varepsilon_{yz} \\ \varepsilon_{xz} \\ \varepsilon_{xy} \end{Bmatrix} =
\begin{bmatrix}
C_{11} & C_{12} & C_{12} & 0 & 0 & 0 \\
C_{12} & C_{11} & C_{12} & 0 & 0 & 0 \\
C_{12} & C_{12} & C_{11} & 0 & 0 & 0 \\
0 & 0 & 0 & C_{44} & 0 & 0 \\
0 & 0 & 0 & 0 & C_{44} & 0 \\
0 & 0 & 0 & 0 & 0 & C_{44}
\end{bmatrix}
\begin{Bmatrix} \sigma_{xx} \\ \sigma_{yy} \\ \sigma_{zz} \\ \sigma_{yz} \\ \sigma_{xz} \\ \sigma_{xy} \end{Bmatrix} \tag{2.16}
$$

Thus, for isotropic linear elastic materials, there are at most three unique material constants to be measured experimentally—C_{11}, C_{12}, and C_{44}. These can be related to Young's modulus and Poisson's ratio as follows. First note that in the uniaxial test performed in Fig. 2.11 the stress state is uniaxial, i.e., $\sigma_{xx} \neq 0, \sigma_{yy} = \sigma_{zz} = \sigma_{xy} = \sigma_{xz} = \sigma_{yz} = 0$. Substituting this into (2.16) results in the following:

$$\varepsilon_{xx} = C_{11}\sigma_{xx} \tag{2.17}$$

$$\varepsilon_{yy} = \varepsilon_{zz} = C_{12}\sigma_{xx} \tag{2.18}$$

Comparing (2.12) and (2.17) reveals the following equivalence

$$C_{11} = \frac{1}{E} \tag{2.19}$$

Comparing (2.13) and (2.18) and using (2.17) and (2.19) results in

$$C_{12} = -\frac{v}{E} \tag{2.20}$$

Thus, two of the three unknown coefficients in the C matrix for an isotropic linear elastic material can be obtained from a uniaxial test. It will now be shown that the third coefficient, C_{44}, may also be obtained from these same coefficients. This can be accomplished by introducing a new experiment, called a **pure shear test**, as shown in Fig. 2.12. In this test, similar to the uniaxial bar test described above, the stress may be plotted versus the applied strain, with the result that for a linear elastic material:

$$\sigma_{xy} = G\varepsilon_{xy} \tag{2.21}$$

where the slope of the curve, G, is called the shear modulus. The yield strength in this test is denoted by σ^S. By equating (2.16) and (2.21) it is apparent that for an isotropic material

$$C_{44} = C_{55} = C_{66} = \frac{1}{G} \tag{2.22}$$

It will now be shown that G is redundant for isotropic media. To do this, consider once again the shear test in Fig. 2.13. Given the shear stress $\sigma_{xy} = k$, where k is an arbitrary constant value of stress applied in the shear test, a cutting plane passed through the stress block in the shear test will result in **free body diagram A**, as shown in the lower left portion of the figure. Summing forces in the x' coordinate direction on this diagram will result in

$$\sum F_{x'} = 0 = \sigma_{y'x'}A + \sigma_{xy}A \sin 45° \cos 45° - \sigma_{xy}A \sin 45° \cos 45° \Rightarrow \sigma_{y'x'} = 0 \tag{2.23}$$

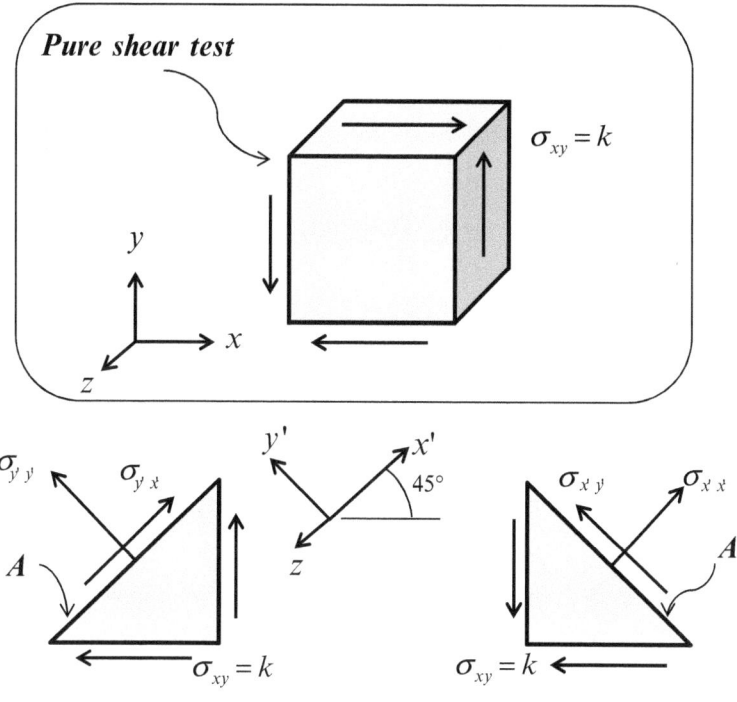

Free Body Diagram A   **Free Body Diagram B**

Fig. 2.13   Pure shear test

Summing forces in the

$$\sum F_{y'} = 0 = \sigma_{y'y'}A + \sigma_{xy}\cos^2 45° + \sigma_{xy}\sin^2 45° \Rightarrow \sigma_{y'y'} = -\sigma_{xy} = -k \quad (2.24)$$

Similarly, summing forces in the x' and y' coordinate directions in *free body diagram B* will result in the following:

$$\sigma_{x'x'} = k \quad (2.25)$$

and

$$\sigma_{x'y'} = 0 \quad (2.26)$$

Thus, it can be seen that the stress states shown in Fig. 2.14 are mechanically equivalent.

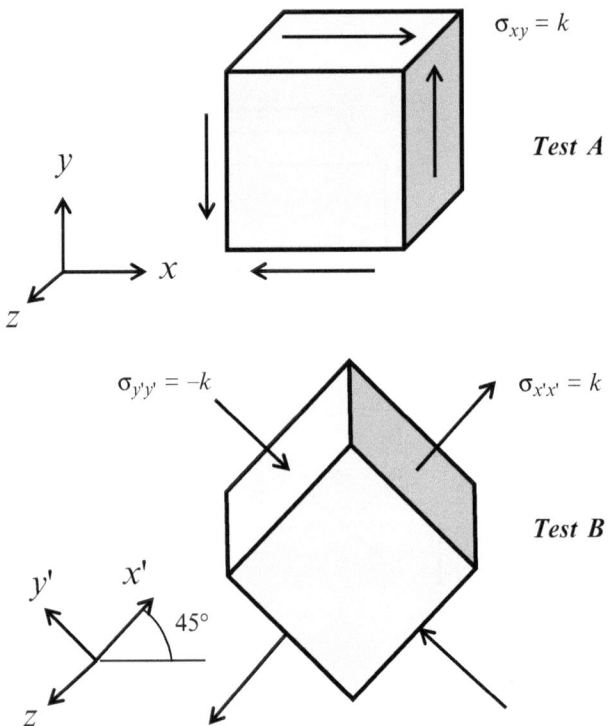

Fig. 2.14 Two mechanically equivalent tests

It is not actually necessary to perform these two tests. Rather, they are performed as a thought experiment in order to relate G to E and v. In order to accomplish this, recall that the material is assumed to be isotropic, so that the stress–strain equations described in (2.16) apply both in the primed and in the unprimed coordinate systems. Applying Test B in Fig. 2.14 to (2.16) and employing (2.19) and (2.20) thus results in the following:

$$\varepsilon_{x'x'} = \frac{(1+v)}{E}k \qquad (2.27)$$

But since Tests A and B are mechanically equivalent, it follows that $\varepsilon_{x'x'} = \varepsilon_{xy}$, and (2.21) and (2.27) can be equated, with the result that

$$G = \frac{E}{(1+v)} \qquad (2.28)$$

Thus, there are only two unique material constants for linear elastic materials, and both constants may be obtained from a uniaxial test: E and v. Therefore, substituting (2.19), (2.20), and (2.28) into (2.16) results in the following:

$$
\begin{Bmatrix} \varepsilon_{xx} \\ \varepsilon_{yy} \\ \varepsilon_{zz} \\ \varepsilon_{yz} \\ \varepsilon_{xz} \\ \varepsilon_{xy} \end{Bmatrix} = \frac{1}{E} \begin{bmatrix} 1 & -v & -v & 0 & 0 & 0 \\ -v & 1 & -v & 0 & 0 & 0 \\ -v & -v & 1 & 0 & 0 & 0 \\ 0 & 0 & 0 & 1+v & 0 & 0 \\ 0 & 0 & 0 & 0 & 1+v & 0 \\ 0 & 0 & 0 & 0 & 0 & 1+v \end{bmatrix} \begin{Bmatrix} \sigma_{xx} \\ \sigma_{yy} \\ \sigma_{zz} \\ \sigma_{yz} \\ \sigma_{xz} \\ \sigma_{xy} \end{Bmatrix}
\qquad (2.29)
$$

As a small footnote, the reader may be interested to know that when the three-dimensional theory of isotropic linear elastic media was first reported in 1822 by Augustin Cauchy, it was thought that there was only a single material constant necessary to describe the constitution. A small but spirited controversy erupted over whether there was one or two material constants for isotropic media, as a fair proportion of the scientific community thought there to be only one. Those supporting the position that there was but a single constant held the view that Poisson's ratio was the same for all materials and should therefore not be included. The controversy seems to have raged almost all the way to the end of the nineteenth century, finally settling on the now well-known correct value of two material constants with the publication of A.E.H. Love's two volume text in 1892–1893 (Love 1892). Perhaps this controversy can be attributed to the fact that in the nineteenth century the state of experimental laboratories was such that it was not possible to measure Poisson's ratio to even one significant digit accuracy. Fortunately, a broad range of experimental techniques were developed in the twentieth century, and today we have accurate values for Poisson's ratio for essentially all materials.

This completes the description of the constitutive behavior of isotropic linear elastic media. A table of material properties for typical engineering materials is given in the appendix.

2.3   Units of Measure

The choice of a system of units for the purpose of this course is not a trivial one. On the one hand, we have the US (or English) system that is commonly used in this country. On the other hand, we have the SI (or metric) system, short for Système International, which is commonly used everywhere else in the world today (except Myanmar!).

In order to explain this strange divergence of systems, it is perhaps necessary to go all the way back to the year 1066, when the French nobleman William of Normandy defeated the English king Harold at Hastings and became the King of England. As a result of this famous battle and the complex system of choosing nobility in Europe, there ensued major differences of opinion between the French and the English that have in some ways continued down to the present. Chief among these were the Hundred Years War and the War of the Roses, but smaller and yet still significant differences of opinion have pervaded Western European culture as well.

For example, when in 1582 Pope Gregory decided to correct the Julian calendar, which had by that time become badly out of time with the sun, the French adopted the Gregorian calendar, but the English refused to accept it until 1752, with the dismaying result that the English Channel divided two countries not only by water, but also by at least 10 days on the calendar.

Such differences may seem overwrought today, but 300 years ago they were all too common. In 1707 a British fleet was destroyed, along with 1,400 sailors, off the Isles of Scilly because it was not possible to measure longitude accurately at the time. Thus, a prize was set by the English monarchy for the first person who could measure longitude accurately. This prize was eventually claimed by John Harrison (in 1767), who invented the ship's chronometer, a clock that is capable of measuring time accurately on a ship at sea. His four clocks, developed over a 31 year span, are still displayed today at the Royal Observatory in Greenwich, east of London (see Fig. 2.15). It is for this reason that the English are credited with establishing the units of time, and the Prime (0°) Meridian is universally accepted to be at the Greenwich Observatory.

The French were not satisfied with this development by the English. Thus, they set out to construct a universally accepted measure of distance. This was accomplished by two French surveyors, Monsieurs Jean Baptiste Joseph Delambre and Pierre Méchain during the 1790s. They were commissioned by the French Academy of Sciences to survey from Barcelona to the Pas de Calais so that the distance from the North Pole to the equator could be accurately determined. The meter was subsequently defined to be on ten-millionth of that distance, and a platinum bar was constructed as a reference, as shown in Fig. 2.16.

Today, the meter is accepted almost everywhere on Earth as the standard unit of distance. Therefore, in this textbook this unit of distance will be adopted, along with the SI system in its entirety. Accordingly, a unit of mass, called a gram, as it was originally conceived was defined to be the mass of 1 dm^3 of water at 0°C. This has now been altered (without destroying the original spirit of the definition) to the mass of a physical prototype preserved by the International Bureau of Weights and Measures.

An important feature of the SI system of units is that there is a clear difference between force and mass, which is often not the case in the US system. In the SI system, the distinction is given by the following formula:

$$F = mg^{\mathrm{E}} \tag{2.30}$$

where force, F, is measured in Newtons (N), m is mass, measured in grams (g), and g^{E} is Earth's gravitational constant, given by

$$g^{\mathrm{E}} = \frac{9.80665\,\mathrm{N}}{10^3\,\mathrm{g}} \tag{2.31}$$

Fig. 2.15 Photos of John Harrison's chronometers, numbers one through four clockwise from *top left*

Table 2.1 shows a table demonstrating the important units of measure in the SI system.

Table 2.2 shows a table demonstrating the important units of measure in the US system.

Table 2.3 is useful for converting from one system to the other. In keeping with the necessity for three significant digits accuracy, conversions are shown to four significant digits accuracy.

Fig. 2.16 Photo of platinum bar representing the meter in the Musée des Artes et Metiers, Paris

Table 2.1 SI units of measure in mechanics

Unit of measure	SI units	Other SI units	Other SI units	Other SI units
Mass	Gram (g)	Kilogram = 10^3 g	mg = 10^{-3} g	
Force	Newton (N)	kN = 10^3 N		
Length	Meter (m)	km = 10^3 m	mm = 10^{-3} m	micron (μm) = 10^{-6}m
Stress	Pascal ($Pa = N/M^2$)	KPa = 10^3 Pa	MPa = 10^6 Pa	GPa = 10^9 Pa
Moment	N m			
Density	g/m^3	kg/m^3		

Table 2.2 US units of measure in mechanics

Unit of measure	US units	Other US units	Other US units	Other US units
Mass	Pound (lb)	Ton = 2×10^3 lb	Ounce(oz) = lb/16	
Force	Pound force (lbf)			
Length	Inch (in.)	Foot (ft) = 12 in.	Yard (yd) = 3 ft	Mile (mi) = 5,280 ft
Stress	lbf/in.^{2} (psi)	lbf/ft^2(psf)	ksi = 10^3 psi	Msi = 10^6 psi
Moment	ft-lbf			
Density	lb/in.^{3}	lb/ft^3		

Table 2.3 Table for converting from SI to US system (to convert to the units at top multiply the value in the column on the left by the factor in the table)

To	lb	lbf	in.	ft	yd	mi	psi	psf	ksi	ft-lbf	lb/in.²	lb/ft²
From												
g	2.205×10^{-3}											
kg	2.205											
mg	2.205×10^{-4}											
N		0.2248										
kN		0.2248×10^{1}										
m			39.37	3.281	1.094	0.6214×10^{-2}						
km			39.37×10^{1}	3.281×10^{2}	1.094×10^{2}	0.6214						
mm			39.37×10^{-2}	3.281×10^{-3}	1.094×10^{-3}	0.6214×10^{-6}						
μm			39.37×10^{-4}	3.281×10^{-4}	1.094×10^{-4}	0.6214×10^{-5}						
Pa							0.1450×10^{-3}	0.02089	0.1450×10^{-4}			
kPa							0.145	20.89	0.1450×10^{-3}			
MPa							0.145×10^{3}	20.89×10^{2}	0.145			
Gpa							0.145×10^{6}	20.89×10^{4}	0.145×10^{6}			
N-m										0.7376		
g/m²											0.036	62.43

2.4   Assignments

PROBLEM 2.1

GIVEN: A uniaxial bar of length, L, is prismatic (meaning the cross-section in the y–z plane does not vary in the x direction), with cross-sectional area, A, and is subjected to an axial load, F_x, at the free end.

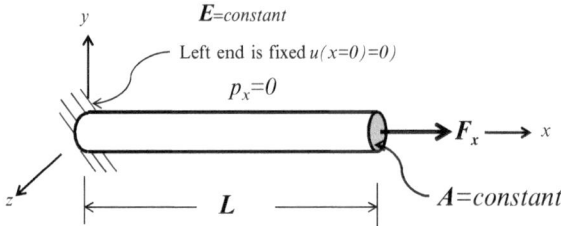

REQUIRED: Consider the axial displacement, $u(x = L)$, of the free end

(a) Propose based on your intuition whether $u(x = L) \propto F$ or $u(x = L) \propto 1/F$ (where $\propto$ means "proportional to"), assuming L and A are held fixed. Plot a graph of $u(x = L)$ vs. F.
(b) Propose based on your intuition whether $u(x = L) \propto L$ or $u(x = L) \propto 1/L$, assuming F and A are held fixed. Plot a graph of $u(x = L)$ vs. F.
(c) Propose based on your intuition whether $u(x = L) \propto A$ or $u(x = L) \propto 1/A$, assuming F and L are held fixed. Plot a graph of $u(x = L)$ vs. A.
(d) Based on your results obtained in parts (a)–(c), propose a general formula for $u(x = L)$ as a function of $F, L,$ and A.

PROBLEM 2.2

GIVEN: The Hoover damn shown below is subjected to a loading that produces stress in the object at all points.

Photo of Hoover Dam Releasing Water (courtesy Bureau of Reclamation PD-USGOV)

The stress state at the midpoint bottom of the damn is given by

$$
\begin{bmatrix}
\sigma_{xx} & \sigma_{xy} & \sigma_{xz} \\
\sigma_{yx} & \sigma_{yy} & \sigma_{yz} \\
\sigma_{zx} & \sigma_{zy} & \sigma_{zz}
\end{bmatrix}
=
\begin{bmatrix}
-200 & 20 & 10 \\
20 & -50 & -25 \\
10 & -25 & -10
\end{bmatrix}
$$

REQUIRED: Draw a depiction of the material point (including coordinate axes) with the stress components labeled on the faces of the resulting cube.

PROBLEM 2.3
GIVEN: The object shown below is subjected to the loads shown. Assume that the object is in equilibrium and that the stresses on the bottom surface of the object are evenly distributed.

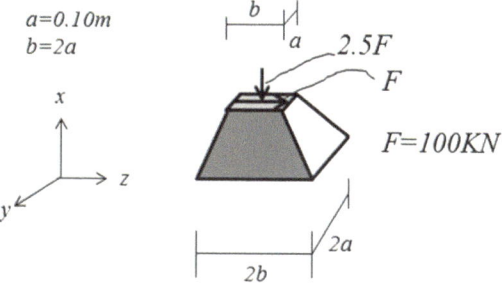

REQUIRED: Determine all nine components of the stress on the bottom surface of the object and draw these components on the stress cube, showing the coordinate axes.

PROBLEM 2.4

GIVEN: The definition of a creep and recovery test is to take a prismatic bar of homogeneous material and apply a constant load in the axial direction, measuring the deformation as a function of time. After a period of time, the load is removed, and the deformation is measured for a length of time equivalent to that over which the load was applied.

REQUIRED

(1) Select a person from this class and work in a team of two, turn in your report together, and try to make it clear, concise, and professional.

(2) Go on the web and find a recipe for making either silly putty or gak; use this recipe to make your own batch of either.

(3) Use your batch to make a cylindrical prismatic bar; measure the length and cross-sectional area of the specimen and report them.

(4) Select a gage length within the specimen and draw dots on the bar at the upper and lower points of the gage length.

(5) Devise a means of holding up the specimen with the long axis aligned vertically, and apply a load axially at the bottom of the specimen in tension in accordance with the given above; report the weight of the applied load; take a photo and include in your report.

(6) Measure the deformation within the gage length at selected time intervals and record them in an excel spreadsheet.

(7) After a selected interval of time (during which the bar does not fracture), remove the load and continue to record the deformation as a function of time in accordance with the given above.

(8) Using the results of (3)–(7), calculate the applied stress as a function of time, as well as the strain as a function of time, and record them in your spreadsheet; turn in your spreadsheet with your report.

(9) Plot graphs of the stress and strain versus time and include in your report.

PROBLEM 2.5

GIVEN: Two identical uniaxial bars with cross-sectional area 0.01 m^2 are subjected to loadings in the x direction at two different loading rates $(dF/dt)_1 = 50$ kN/s (Test #1), $(dF/dt)_2 = 100$ kN/s (Test #2)

Time (s)	ε_{xx} (Test #1)	ε_{xx} (Test #2)
0	0	0
5	0.0008	0.0012
10	0.0012	0.0020
15	0.0017	0.0030
20	0.0020	0.0040
25	0.0025	0.0065
30	0.0030	0.0200

(continued)

continued Time (s)	ε_{xx} (Test #1)	ε_{xx} (Test #2)
35	0.0035	0.0630
40	0.0040	X
45	0.0045	
50	0.0095	
55	0.0270	
60	0.0580	
65	X	

X means "fracture"

REQUIRED

(1) Plot a single input diagram showing σ_{xx} versus time for both tests.
(2) Plot a single output diagram showing ε_{xx} versus time for both tests.
(3) Plot a single crossplot diagram of σ_{xx} versus ε_{xx} for both tests.

PROBLEM 2.6

GIVEN: Stress may be expressed in either the metric system (Pascals, meaning Newtons per square meter) or the English system (*psi*, meaning *lbf* per square inch)

REQUIRED

1. Derive the relation between *Pa* and *psi*
2. Derive the relation between *MPa* and *ksi*
3. Derive the relation between *ksi* and *GPa*

References

Allen D, Haisler W (1985) Introduction to aerospace structural analysis. Wiley, New York
Glover C, Jones H (1992) Conservation principles for continuous media. McGraw-Hill, New York
Love AEH (1892) Treatise on the mathematical theory of elasticity. Dover, New York
Malvern L (1969) Introduction to the mechanics of a continuous medium. Prentice-Hall, Englewood Cliffs
Timoshenko S (1953) History of strength of materials. McGraw-Hill, New York

Chapter 3
Theory of Uniaxial Bars

3.1 Introduction

A bar is defined as an object that has one dimension that is large compared to the other two[1]. *If the bar is loaded uniquely in the direction of its long dimension, we call it a uniaxial bar,* and in this case we assign the x coordinate axis to be in the direction of the long dimension of the bar, as shown in Fig. 3.1 (Allen and Haisler 1985).

If the bar is loaded in tension, it is sometimes termed a cable or rope. If the bar is loaded in compression, it is sometimes called a column. An example of a column is shown in Fig. 3.2.

Note that the shape of the cross-sectional area, A, of the bar is not necessarily circular, nor is the bar prismatic, meaning that, in general $A = A(x)$. Furthermore, we will assume that *the bar may be heterogeneous, but the properties do not change in the y or z coordinate directions.* Thus, in general, $E = E(x)$, $\sigma^T = \sigma^T(x), \sigma^C = \sigma^C(x),$ *and* $\sigma^S = \sigma^S(x)$, where E is Young's modulus, and σ^T, σ^C, and σ^S are the yield stresses of the material in tension (T), compression (C), and shear (S), as described in Chap. 2. Note also that the geometry of the bar is specified by the length, L, and the cross-sectional area, A. The external loading is described by the axial loading per unit length, p_x (which may include gravitational loads), and point loads, F_x, that may be applied at points along the length of the bar as well as at the ends of the bar $(x = 0, L)$.

Note that when a bar is subjected to axial loading such as that shown in Fig. 3.1, the bar will necessarily respond to that load by extending along its length. Thus, it should be apparent that for loads applied in the x direction, there will be a component of deformation of the beam, $u = u(x, y, z)$, in the x coordinate direction, and this component is called the axial deflection. In this chapter, we will concern ourselves with the axial deflection of the centroidal axis, termed $u_0 \equiv u(x, \; y = 0, \; z = 0) = u_0(x)$ for reasons that will become clear below.

---

[1] Note: any italicized statement in the text in this section constitutes a model assumption.

D.H. Allen, *Introduction to the Mechanics of Deformable Solids: Bars and Beams*,
DOI 10.1007/978-1-4614-4003-1_3, © Springer Science+Business Media New York 2013

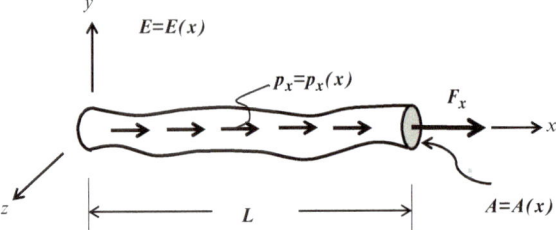

Fig. 3.1 General depiction of a uniaxial bar subjected to mechanical loading

Fig. 3.2 Egyptian obelisk at the temple of Luxor

Table 3.1 Boundary conditions applied to a uniaxial bar

Boundary conditions
(a) On the end $x = 0$, either $F(x = 0) = $ known or $u_0(x = 0) = $ known
(b) On the end $x = L$, either $F(x = L) = $ known or $u_0(x = L) = $ known

The boundary conditions applied on the ends of the bar may be of two types: either loads or displacements. Thus, at each end there is one boundary condition that must be known a priori. These are listed in Table 3.1.

As a review then, all of the inputs required to completely define the problem are described in Table 3.2.

Table 3.2 Uniaxial bar problem inputs

Problem inputs
1. **Loads**
(a) $p_x = p_x(x)$, $F_x(x = x_1)$
(b) Boundary conditions: $F_x(x = 0)$ and $F_x(x = L)$[a]
2. **Geometry**
(a) $A = A(x)$
(b) L
(c) Boundary conditions $u_0(x = 0)$ and $u_0(x = L)$[a]
3. **Material properties**
(a) $E = E(x)$
(b) $\sigma^T = \sigma^T(x)$, $\sigma^C = \sigma^C(x)$, $\sigma^S = \sigma^S(x)$

[a]Note that **either** load **or** displacement is specified at each end

3.2   A Model for Predicting the Mechanical Response of a Uniaxial Bar

For purposes of creating a robust model that can be utilized to avoid failure of the bar due to fracture or excessive deformation, it is recognized that the following output variables need to be predicted: the stress, σ_{xx}, and the axial displacement, u_0. It will be shown that there are two additional outputs necessary to obtain the above two, and these are: the axial internal load, P [to be defined below in (3.3)], and the strain, ε_{xx}. It will also be shown that the assumptions required in order to construct a simple yet accurate model will require that the above four variables vary only in the x coordinate direction. Thus, the problem outputs are summarized in Table 3.3.

3.2.1   Construction of the Model

As can be seen from the above listing of outputs, there are four unknowns in the problem. Therefore, **it is evident that we will need four equations in order to construct a rigorous model**. These are as follows:

1. **Newton's Laws**

$$\sum \vec{F} = 0, \quad \sum \vec{M} = 0 \quad \text{assuming the bar is at rest} \tag{3.1}$$

Note that the above simplifies to the following single equation due to the absence of forces in the y and z directions, and moments about any of the axes:

$$\sum F_x = 0 \tag{3.2}$$

Table 3.3 Uniaxial bar problem outputs

Problem outputs
1. Internal load: $P = P(x)$
2. Axial stress: $\sigma_{xx} = \sigma_{xx}(x)$
3. Axial strain: $\varepsilon_{xx} = \varepsilon_{xx}(x)$
4. Axial displacement: $u_0 = u_0(x)$

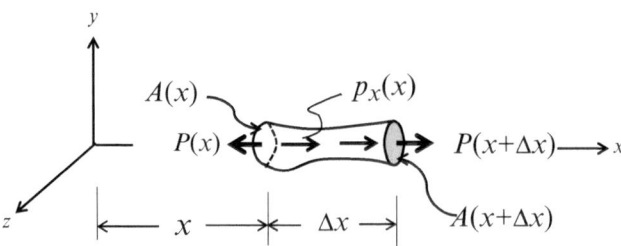

Fig. 3.3 Free body diagram of a section of a uniaxial bar

Now, let us introduce a new variable, P, called the internal axial load and defined as follows:

$$P \equiv \iint \sigma_{xx}\, dy\, dz \tag{3.3}$$

Note that as a consequence of definition (3.3) $P = P(x)$. Next, suppose that a free body diagram of the bar is constructed, depicting two planes passed through the bar normal to the x-axis at coordinate locations x and $x + \Delta x$, as shown in Fig. 3.3.

Equation (3.2) above may now be used to sum forces in the x direction for the section of the bar shown in Fig. 3.3 as follows:

$$\sum F_x = 0 \Rightarrow P(x + \Delta x) - P(x) + \int_{x}^{x+\Delta x} p_x(x)\, dx \tag{3.4}$$

Dividing the above equation through by Δx and invoking the fundamental theorem of calculus, as well as the definition of an ordinary derivative will result in the following equilibrium equation for the uniaxial bar.

$$\frac{dP}{dx} = -p_x(x) \tag{3.5}$$

2. Kinematics

(a) Strain–displacement relation

$$\varepsilon_{xx} \equiv \frac{\partial u}{\partial x} \tag{3.6}$$

(b) Kinematic assumption—*cross-sections that are planar and normal to the x-axis before loading remain planar and normal to the x-axis after loading.* The consequence of this assumption is that the axial displacement, u, is not a function of y or z. Therefore, $u(x, y, z) = u_0(x)$, and we will drop the subscript zero in the remainder of this chapter. It also follows from (3.6) that ε_{xx} also is not a function of y and z, so that $\varepsilon_{xx} = \varepsilon_{xx}(x)$.

(c) Constitutive equation—A consequence of the assumption that the loads are applied axially, together with kinematic assumption (b) above is that

$$\sigma_{xx} \gg \sigma_{yy}, \sigma_{zz} \Rightarrow \sigma_{yy}, \sigma_{zz} \approx 0 \tag{3.7}$$

Thus, assuming *that the material is orthotropic and behaves linear elastically at all points in the bar*

$$\varepsilon_{xx} = \frac{\sigma_{xx}}{E} \Leftrightarrow \sigma_{xx} = E\varepsilon_{xx} \tag{3.8}$$

Substituting (3.6) into (3.8) results in the following

$$\sigma_{xx} = E\frac{du}{dx} \tag{3.9}$$

Since it is clear from the above that $\sigma_{xx} = \sigma_{xx}(x)$, it follows that substituting (3.9) into (3.3) results in

$$P = EA\frac{du}{dx} \tag{3.10}$$

Combining (3.9) and (3.10) thus results in

$$P = P(x) = \sigma_{xx}(x)A(x) \Leftrightarrow \sigma_{xx} = P/A \tag{3.11}$$

Now, let us very carefully examine (3.5), (3.6), (3.8), and (3.11). By comparing these equations to information listed in Tables 3.2 and 3.3, it can be seen that there are three kinds of variables listed in these four equations: inputs (*which are known!*), outputs (*which are to be determined!*), and the independent variable, x, which the outputs are to be determined as functions of. For convenience, the four equations are listed in Table 3.4 with the inputs circled and the outputs in boxes.

The above set of equations, together with the boundary conditions described in Table 3.1, constitute what is known as "a well-posed boundary value problem." That is due to the fact that there are four equations in four unknowns: P, σ_{xx}, ε_{xx}, and u. In addition, there are exactly two derivatives in the equations, thus requiring two boundary conditions, and finally, it can be shown that all four of the equations are mathematically linear, so that one can prove that the above set of equations has a unique solution. ***Thus, we have a mathematically acceptable model, but it remains to be seen if it is a physically accurate model.***

Table 3.4 Governing equations for uniaxial bar model

$$\boxed{\frac{dP}{dx}} = -\boxed{p_x}(x) \qquad (3.5)$$

$$\boxed{\varepsilon_{xx}} \equiv \boxed{\frac{du}{dx}} \qquad (3.6)$$

$$\boxed{\varepsilon_{xx}} = \boxed{\sigma_{xx}}/\boxed{E} \qquad (3.8)$$

$$\boxed{\sigma_{xx}} = \boxed{P}/\boxed{A} \qquad (3.11)$$

For convenience, let us review the assumptions we made in order to construct our model. This is important because if we attempt to use the model to design an object for which any of the assumptions are violated, we are going to necessarily introduce some error into our model. These assumptions, which were previously italicized in the text above, are listed in Table 3.5.

Table 3.5 Assumptions used to construct a model for a uniaxial bar subjected to mechanical loading

Assumptions used to construct the model

1. The object has one dimension that is large compared to the other two (called a bar)
2. The bar is loaded uniquely in the direction of its largest dimension (called a uniaxial bar)
3. The material properties do not change in the y or z coordinate directions (normal to the long axis: x)
4. The bar is at rest
5. Cross-sections that are planar and normal to the x-axis before loading remain planar and normal to the x-axis after loading
6. The material is orthotropic and behaves linear elastically at all points in the bar

3.2.2 Methods for Obtaining Solutions with the Model

Unfortunately, we are not quite yet finished with our model. We still need to develop systematic methods for solving the four equations shown in Table 3.4 for the four unknowns: P, σ_{xx}, ε_{xx}, and u. In order to do that, a few observations regarding the equations shown in Table 3.4 are appropriate. These are as follows.

1. Note that the equations are partially coupled, meaning that more than one of the unknowns (in boxes) occurs in several of the equations [except equation (3.5)]!
2. The equations are all ordinary differential equations, meaning that there is only one independent variable: x.
3. A careful examination of the equations reveals that all of our inputs occur explicitly in the equations in the form of loads: p_x and P, geometry: A (and L, via the boundary conditions), and material property: E.

4. As a consequence of observation (3), the resulting four predictive equations for P, σ_{xx}, ε_{xx}, and u that we will obtain by solving the four equations in Table 3.4 will turn out to contain the inputs $(p_x, A, L, and E)$ as well as x on the right hand side of the four equations. **When we have these equations, we will have a model that we can use to design uniaxial bars.**

Using the above observations, let us now proceed to construct systematic methods for obtaining equations of the form described in observation (4). As it turns out, there are lots of ways of solving the set of four equations. Sometimes one way is easier than another, and this creates lots of confusion for students. We want to minimize this confusion, so we will develop consistent approaches that will always work.

3.2.2.1   Solution Methods for Statically Determinate Uniaxial Bars

Recall that either a force or a displacement boundary condition is applied at each end of the bar but not both. There must be at least one displacement boundary condition in order for the solution to be unique. In the case where a displacement boundary condition is applied at one end of the bar, and a force boundary condition is applied at the other end, the problem is called "statically determinate." That is due to the fact that in this case one can construct a free body diagram of the entire bar and determine the reaction where the displacement boundary condition is applied by employing equilibrium equation (3.2). The solution method for statically determinate uniaxial bars is described in Table 3.6.

Table 3.6 Systematic method for solving the unknowns in a statically determinate uniaxial bar subjected to mechanical loading

Systematic solution method for statically determinate uniaxial bars
Step 1: Solve (3.5) for $P = P(x)$ using direct integration (and one force boundary condition)
Step 2: Use $P(x)$ obtained in step 1 to obtain $\sigma_{xx} = \sigma_{xx}(x)$ using (3.11)
Step 3: Use $\sigma_{xx}(x)$ obtained in step 2 to obtain $\varepsilon_{xx} = \varepsilon_{xx}(x)$ in (3.8)
Step 4: Use $\varepsilon_{xx}(x)$ obtained in step 3 to obtain $u = u(x)$ using direct integration in (3.6) (and one displacement boundary condition)

The above method will result in four explicit equations for the unknowns P, σ_{xx}, ε_{xx}, and u as functions of the inputs $p_x, A, L, and E$ and the independent variable, x. Once these equations are constructed, they can be used to ensure that the bar is designed in such a way that the design constraints (in this course, fracture and excessive displacements) can be satisfied.

There is one remaining question to be answered: is this model accurate? The answer to this question can only be determined by performing systematic experiments that can be used to test the validity of the model. Fortunately, many experiments have been performed to test this theory (beginning in the eighteenth century), and it is generally agreed upon by the scientific and engineering communities that the model is accurate so long as the assumptions listed in

Table 3.5 are not violated. Should the designer using this model use the model to design a uniaxial bar wherein any of the model assumptions are violated, it is recommended that the model be supported by experiments designed to test the accuracy of the model with respect to the assumptions that have been violated in the design process.

Example Problem 3.1
Given: The prismatic and homogeneous uniaxial bar shown below has a force F applied at the free end $(x = L)$.

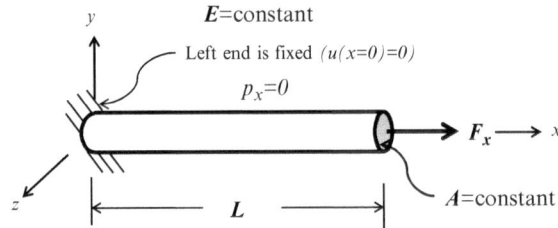

Required

1. Using uniaxial bar theory, derive an expression for each of the following:

   (a) $P = P(x, F_x)$
   (b) $\sigma_{xx} = \sigma_{xx}(x, F_x, A, L, E)$
   (c) $\varepsilon_{xx} = \varepsilon_{xx}(x, F_x, A, L, E)$
   (d) $u = u(x, F_x, A, L, E)$

2. Plot the results of (a)–(d) on four different graphs: $P = P(x)$, $\sigma_{xx} = \sigma_{xx}(x)$ $\varepsilon_{xx} = \varepsilon_{xx}(x)$, and $u = u(x)$, and

3. Suppose the bar is 10 m long, with $F_x = 100,000$ kN and is made of A36 steel $(E = 200$ GPa$)$, and the maximum allowable displacement of the right end of the bar is 0.01 m, determine the minimum allowable cross-sectional area of the bar, A

Solution

1. (a) since $p_x = 0$, it follows that, from equation (3.5)

$$\frac{dP}{dx} = -p_x = 0 \Rightarrow \int \frac{dP}{dx}\,dx = \int 0\,dx \Rightarrow P(x) = C_1 \qquad (E3.1.1)$$

Now consider the force, F_x, applied at the right end of the bar. Newton's third law may be used to deduce that the boundary condition at the right end is $P(x = L) = F_x$. Applying this boundary condition to (E3.1.1) results in

$$\boxed{P(x) = F_x} \quad \text{Q.E.D.} \qquad (E3.1.2)$$

(b) Using (3.7) and (E3.1.2) results in the following

$$\boxed{\sigma_{xx} = \frac{P}{A}} \Rightarrow \sigma_{xx} = \frac{F_x}{A} \quad \text{Q.E.D.} \tag{E3.1.3}$$

(c) Using (3.8) and (E3.1.3) results in the following

$$\varepsilon_{xx} = \frac{\sigma_{xx}}{E} \Rightarrow \boxed{\varepsilon_{xx} = \frac{F_x}{EA}} \quad \text{Q.E.D.} \tag{E3.1.4}$$

(d) Using (3.6) and (E3.1.4) results in the following

$$\frac{du}{dx} = \varepsilon_{xx} = \frac{F_x}{EA} \Rightarrow \int \frac{du}{dx}\,dx = \int \frac{F_x}{EA}\,dx \Rightarrow u(x) = \frac{F_x x}{EA} + C_2 \tag{E3.1.5}$$

Applying the boundary condition $u(x = 0) = 0$ to (E3.1.5) implies that $C_2 = 0$. Therefore, (E3.1.5) simplifies to the following

$$\boxed{u(x) = \frac{F_x x}{EA}} \quad \text{Q.E.D.} \tag{E3.1.6}$$

2. Plotting (E3.1.2)–(E3.1.4), and (E3.1.6) results in

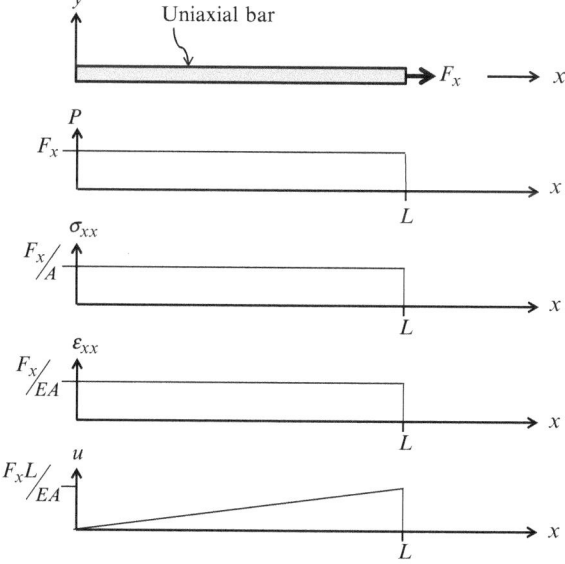

3. The maximum displacement occurs at $x = L$, so that (E3.1.6) gives

$$u_{max} = u(x = L) = \frac{F_x L}{EA} \Rightarrow A_{min} = \frac{F_x L}{E u_{min}} \tag{E3.1.7}$$

Applying the values from the problem given result in

$$A_{min} = \frac{100,000\,\text{kN} \times 10\,\text{m}}{200\,\text{GPa} \times 0.01\,\text{m}} \times \frac{10^3\,\text{N}}{\text{kN}} \times \frac{\text{GPa}}{10^9\,\text{Pa}} \times \frac{\text{Pa}}{\text{N/m}^2} \Rightarrow$$

$$\boxed{A_{min} = 0.5\,\text{m}^2} \quad \text{Q.E.D.} \tag{E3.1.8}$$

3.2.2.2 Solution Methods for Statically Indeterminate Uniaxial Bars

In the case where displacement boundary conditions are applied at both ends of the uniaxial bar, the problem is called "statically indeterminate" because there are two unknown reactions and only one equilibrium equation, thus making it intractable to find the reactions using equilibrium equation (3.2) by itself. In this case an alternate solution method is preferable. The method is constructed by first substituting (3.10) into (3.5), resulting in the following second-order differential equation in u.

$$\frac{d}{dx}\left(EA\frac{du}{dx}\right) = -p_x(x) \tag{3.12}$$

Since there are two displacement boundary conditions in statically indeterminate problems, (3.12) can be solved directly to obtain the displacement, $u = u(x)$. The remaining unknowns can be found by back substitution into the other equations, as described in Table 3.7.

Table 3.7 Systematic method for solving the unknowns in a statically indeterminate uniaxial bar subjected to mechanical loading

Systematic solution method for statically indeterminate uniaxial bars
Step 1: Solve (3.12) for $u = u(x)$ using direct integration (and two displacement boundary conditions)
Step 2: Use $u(x)$ obtained in step 1 to obtain $P = P(x)$ using (3.10)
Step 3: Use $P(x)$ obtained in step 2 to obtain $\sigma_{xx} = \sigma_{xx}(x)$ using (3.11)
Step 4: Use $\sigma_{xx}(x)$ obtained in step 3 to obtain $\varepsilon_{xx} = \varepsilon_{xx}(x)$ using (3.8)

Example Problem 3.2
Given: The prismatic and homogeneous uniaxial bar shown below has a constant applied load per unit length $p_x = p_x^0$.

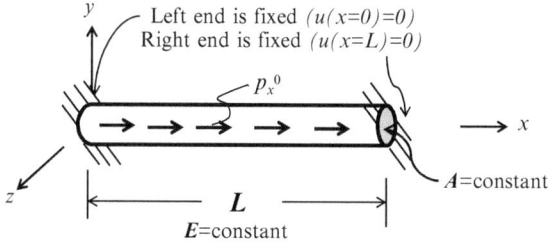

Required

1. Using uniaxial bar theory, derive an expression for each of the following:

(a) $u = u(x, p_x^0, A, L, E)$
(b) $P = P(x, p_x^0, A, L, E)$
(c) $\sigma_{xx} = \sigma_{xx}(x, p_x^0, A, L, E)$
(d) $\varepsilon_{xx} = \varepsilon_{xx}(x, p_x^0, A, L, E)$

2. Plot the results of (a)–(d) on four different graphs: $P = P(x)$, $\sigma_{xx} = \sigma_{xx}(x)$, $\varepsilon_{xx} = \varepsilon_{xx}(x)$, and $u = u(x)$
3. Find the reactions at the left and right ends of the bar

Solution

1. (a) since $p_x^0 = $ constant, it follows that, from equation (3.11)

$$\frac{d}{dx}\left(EA\frac{du}{dx}\right) = -p_x^0 \Rightarrow \int \frac{d}{dx}\left(EA\frac{du}{dx}\right)dx = -\int p_x^0 \, dx \Rightarrow$$

$$EA\frac{du}{dx} = -p_x^0 x + C_1 \Rightarrow \int EA\frac{du}{dx}\,dx = \int \left(-p_x^0 x + C_1\right)dx \Rightarrow$$

$$u(x) = -\frac{p_x^0 x^2}{2EA} + \frac{C_1 x}{EA} + C_2 \tag{E3.2.1}$$

Now consider the boundary condition $u(x = 0) = 0$. It can be seen from (E3.2.1) that

$$C_2 = 0 \tag{E3.2.2}$$

Next consider the boundary condition $u(x = L) = 0$. It can be seen from (E3.2.1) and (E3.2.2) that

$$0 = -\frac{p_x^0 L^2}{2EA} + \frac{C_1 L}{EA} \Rightarrow C_1 = \frac{p_x^0 L}{2} \tag{E3.2.3}$$

Substituting (E3.2.2) and (E3.2.3) into (E3.2.1) thus results in

$$\boxed{u(x) = -\frac{p_x^0 x^2}{2EA} + \frac{p_x^0 L x}{2EA}} \quad \text{Q.E.D.} \tag{E3.2.4}$$

(b) Substituting (E3.2.4) into (3.10) results in the following:

$$P(x) = EA \frac{du}{dx} = EA \frac{d}{dx}\left(-\frac{p_x^0 x^2}{2EA} + \frac{p_x^0 L x}{2EA} \right) \Rightarrow$$

$$\boxed{P(x) = -p_x^0 x + \frac{p_x^0 L}{2}} \quad \text{Q.E.D.} \tag{E3.2.5}$$

(c) Substituting (E3.2.5) into (3.7) results in the following:

$$\boxed{\sigma_{xx}(x) = -\frac{p_x^0 x}{A} + \frac{p_x^0 L}{2A}} \quad \text{Q.E.D.} \tag{E3.2.6}$$

(d) Substituting (E3.2.6) into (3.8) results in the following:

$$\boxed{\varepsilon_{xx}(x) = -\frac{p_x^0 x}{EA} + \frac{p_x^0 L}{2EA}} \quad \text{Q.E.D.} \tag{E3.2.7}$$

2. Plotting (E3.2.4)–(E3.2.7) results in the following:

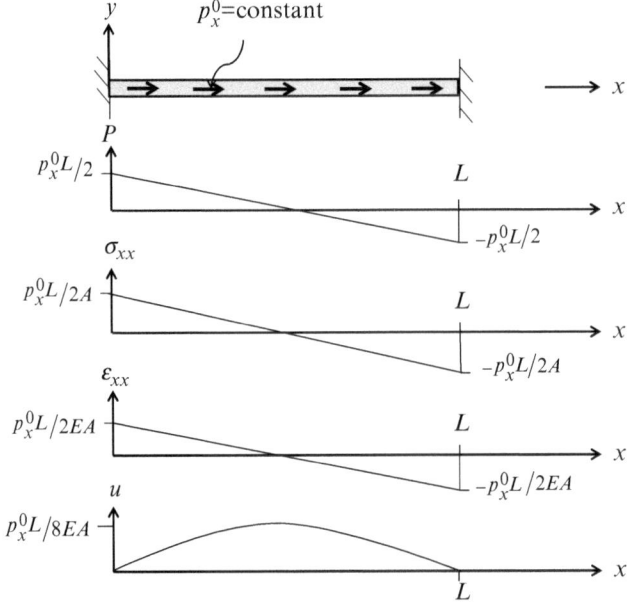

3. It can be seen that the reaction at the left end of the bar, R_L, is equal to $P(x = 0)$. Therefore, using (3.2.5) results in

$$R_L = P(x = 0) = -p_x^0 \times 0 + \frac{p_x^0 L}{2} \Rightarrow \boxed{R_L = \frac{p_x^0 L}{2}} \quad \text{Q.E.D.} \tag{E3.2.8}$$

Also, the reaction at the right end of the bar, R_R, is equal to $P(x = L)$. Therefore, using (E3.2.5) again results in

$$R_R = P(x = L) = -p_x^0 L + \frac{p_x^0 L}{2} \Rightarrow \boxed{R_R = -\frac{p_x^0 L}{2}} \quad \text{Q.E.D.} \quad \text{(E3.2.9)}$$

3.2.2.3   How to Handle Point Forces Applied Within the Bar

Uniaxial bars are sometimes subjected to uniaxial loads that are applied over such short distances in the x coordinate direction that for practical purposes they can be considered to be applied at a single point along the x-axis. We term these loads "point forces." An example would be the resultant axial load applied by the guide wires to a radio or cell phone tower. The inclusion of such forces in our model can be accounted for by performing a careful analysis of the kinetics. To see how this can be done, consider a uniaxial bar with a point force applied to it, as shown in Fig. 3.4.

Now, suppose that the bar is cut normal to the x-axis at coordinate location x_1^+, just to the right of the location of the force, F_x, as shown in Fig. 3.5.

In this case, summing forces will result in the following:

$$\sum F_x = 0 \Rightarrow P(x_1^+) - P(0) + F_x + \int_0^{x_1^+} p_x(x)\, dx = 0 \Rightarrow$$

$$P(x_1^+) = P(0) - F_x - \int_0^{x_1^+} p_x\, dx \tag{3.13}$$

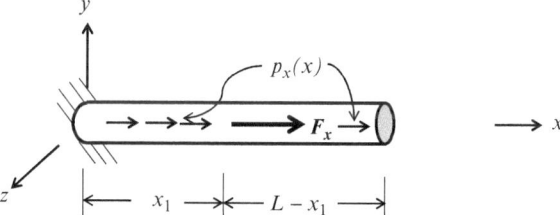

Fig. 3.4 Uniaxial bar subjected to point force, F_x

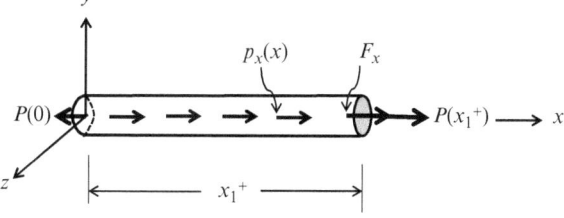

Fig. 3.5 Free body diagram of uniaxial bar cut to the right of load, F_x

On the other hand, if the second cut is made just to the left of where the force F_x is applied, the free body diagram is identical to that shown in Fig. 3.3, with the necessary result that

$$P(x_1^-) = P(0) - \int_0^{x_1^-} p_x \, dx \qquad (3.14)$$

It therefore follows that since the coordinate locations x_1^- and x_1^+ can be made arbitrarily close, there must be a jump discontinuity in $P(x)$ at the coordinate location, x_1. If the force, F_x, is applied in the positive x direction, the jump is negative and equal in magnitude to F_x. If the force, F_x, is applied in the negative x direction, then the jump is positive and equal in magnitude to F_x. These results are summarized in Table 3.8.

Table 3.8 How to handle point forces applied within uniaxial bars

How to handle a point force, F_x, applied at coordinate location $x = x_1$
1. If F_x is in the positive x direction, decrease $P(x_1)$ by F_x
2. If F_x is in the negative x direction, increase $P(x_1)$ by F_x

Example Problem 3.3
Given: The prismatic and homogeneous uniaxial bar shown below has point forces F_x^1 and F_x^2 applied as shown.

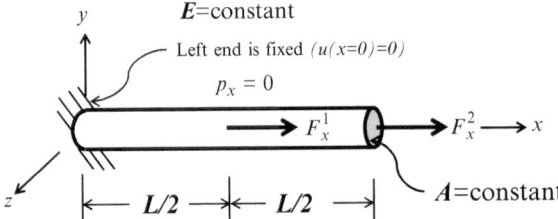

Required

1. Using uniaxial bar theory, derive an expression for each of the following:

   (a) $P = P(x, F_x^1, F_x^2)$
   (b) $\sigma_{xx} = \sigma_{xx}(x, F_x^1, F_x^2, A, L, E)$
   (c) $\varepsilon_{xx} = \varepsilon_{xx}(x, F_x^1, F_x^2, A, L, E)$
   (d) $u = u(x, F_x^1, F_x^2, A, L, E)$

2. Plot the results of (a)–(d) on four different graphs: $P = P(x)$, $\sigma_{xx} = \sigma_{xx}(x)$, $\varepsilon_{xx} = \varepsilon_{xx}(x)$, and $u = u(x)$

Solution

1. Since the problem is statically determinate, use Table 3.6.

   (a) Solve (3.5) for $P = P(x)$ as follows:

$$\frac{dP}{dx} = -p_x = 0 \Rightarrow P = C_1 \quad 0 \le x \le L/2$$
$$P = C_2 \quad L/2 \le x \le L \tag{E3.3.1}$$

   Now apply the boundary condition $P(x = L) = F_x^2$. Thus, using Table 3.8, it follows that

$$\boxed{\begin{aligned} P &= F_x^1 + F_x^2 \quad 0 \le x \le L/2 \\ P &= F_x^2 \quad L/2 \le x \le L \end{aligned}} \quad \text{Q.E.D.} \tag{E3.3.2}$$

   (b) Solve (3.7) for $\sigma_{xx} = \sigma_{xx}(x)$ as follows:

$$\sigma_{xx} = \frac{P}{A} \Rightarrow \boxed{\sigma_{xx} = \frac{F_x^1 + F_x^2}{A} \quad 0 \le x \le L/2}$$
$$\boxed{\sigma_{xx} = \frac{F_x^2}{A} \quad L/2 \le x \le L} \quad \text{Q.E.D.} \tag{E3.3.3}$$

   (c) Solve (3.8) for $\varepsilon_{xx} = \varepsilon_{xx}(x)$ as follows:

$$\varepsilon_{xx} = \frac{\sigma_{xx}}{E} \Rightarrow \boxed{\varepsilon_{xx} = \frac{F_x^1 + F_x^2}{EA} \quad 0 \le x \le L/2}$$
$$\boxed{\varepsilon_{xx} = \frac{F_x^2}{EA} \quad L/2 \le x \le L} \quad \text{Q.E.D.} \tag{E3.3.4}$$

   (d) Solve (3.6) for $u = u(x)$ as follows:

$$\frac{du}{dx} = \varepsilon_{xx} \Rightarrow \frac{du}{dx} = \frac{F_x^1 + F_x^2}{EA} \quad 0 \le x \le L/2 \Rightarrow$$
$$u(x) = \frac{(F_x^1 + F_x^2)}{EA} + C_3 \quad 0 \le x \le L/2$$

   Now apply the boundary condition $u(x = 0) = 0$, from which it is clear that $C_3 = 0$, thus resulting in

$$\boxed{u(x) = \frac{(F_x^1 + F_x^2)x}{EA} \quad 0 \le x \le L/2} \quad \text{Q.E.D.} \tag{E3.3.5}$$

Next, consider the right half of the bar.

$$\frac{du}{dx} = \frac{F_x^2}{EA} \Rightarrow \boxed{u(x) = \frac{F_x^2 x}{EA} + C_4 \quad L/2 \le x \le L} \tag{E3.3.6}$$

Now, it is necessary to match the displacement of the right half of the bar to the left half. To do this, it is necessary to obtain a matching condition for the displacement. This can be obtained by evaluating (E3.3.5) at $x = L/2$ as follows:

$$u(x = L/2) = \frac{\left(F_x^1 + F_x^2\right)L}{2EA} \tag{E3.3.7}$$

The above result may be used as a boundary condition for (E3.3.6) to obtain C_4 as follows:

$$u(x = L/2) = \frac{\left(F_x^1 + F_x^2\right)L}{2EA} = \frac{F_x^2 L}{2EA} + C_4 \Rightarrow C_4 = \frac{F_x^1 L}{2EA} \tag{E3.3.8}$$

Thus, substituting (E3.3.8) into (E3.3.6) results in

$$\boxed{u(x) = \frac{F_x^2 x}{EA} + \frac{F_x^1 L}{2EA} \quad L/2 \le x \le L} \quad \text{Q.E.D.} \tag{E3.3.9}$$

2. Plotting (3.3.2)–(3.3.5) and (3.3.9) results in the following:

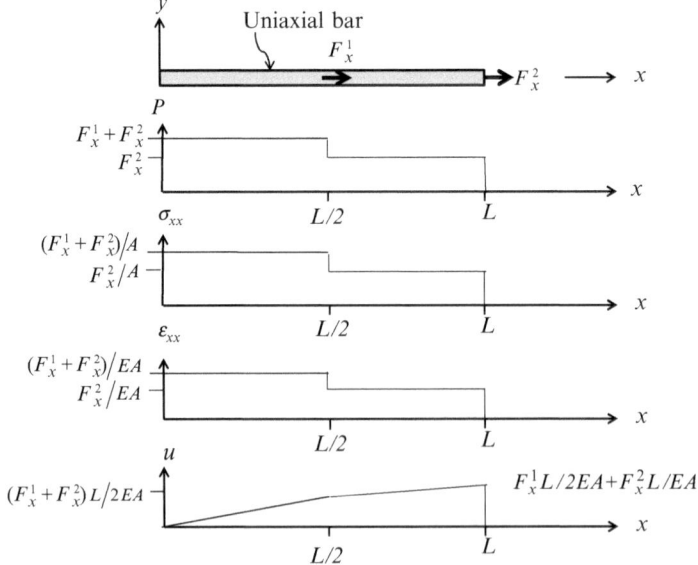

3.3 Assignments

PROBLEM 3.1
GIVEN: The definition of a uniaxial bar (or column).
REQUIRED: Locate a column either on the university campus or in the local area, describe it (meaning loads, geometry, and material properties), and include a photo of it.

PROBLEM 3.2
GIVEN: The uniaxial bar shown below is homogeneous, prismatic and has an evenly distributed axial load of $p_x = p_x^0 =$ constant applied along its length, as well as a point force, $F_x = p_x^0 L$, applied at $x = L/2$, as shown.

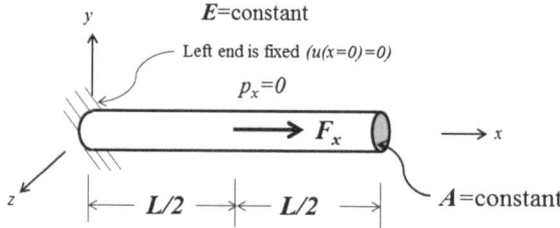

REQUIRED

1. Using uniaxial bar theory, derive an expression for each of the following:

   (a) $P = P(x, F_x, L)$
   (b) $\sigma_{xx} = \sigma_{xx}(x, F_x, A, L, E)$
   (c) $\varepsilon_{xx} = \varepsilon_{xx}(x, F_x, A, L, E)$
   (d) $u = u(x, F_x, A, L, E)$

2. Plot the results of (a)–(d) on four different graphs: $P = P(x)$, $\sigma_{xx} = \sigma_{xx}(x)$, $\varepsilon_{xx} = \varepsilon_{xx}(x)$, and $u = u(x)$ (for a given value of the input loads, geometry, and material properties).
3. Find the location of the maximum stress, σ_{xx} and draw the stress block at that point.

PROBLEM 3.3
GIVEN: The uniaxial bar shown below is homogeneous, prismatic and has an evenly distributed axial load of $p_x = p_x^0 =$ constant applied along its length as shown.

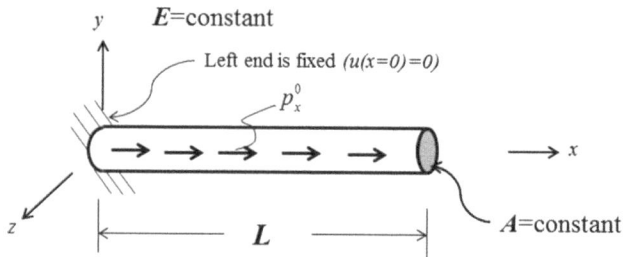

REQUIRED

1. Using uniaxial bar theory, derive an expression for each of the following:

   (a) $P = P(x, p_x^0)$
   (b) $\sigma_{xx} = \sigma_{xx}(x, p_x^0, A, L, E)$
   (c) $\varepsilon_{xx} = \varepsilon_{xx}(x, p_x^0, A, L, E)$
   (d) $u = u(x, p_x^0, A, L, E)$

2. Plot the results of (a)–(d) on four different graphs: $P = P(x)$, $\sigma_{xx} = \sigma_{xx}(x)$, $\varepsilon_{xx} = \varepsilon_{xx}(x)$, and $u = u(x)$ (for a given value of the input loads, geometry, and material properties).

3. Suppose the bar is 10 m long, with $p_x^0 = 10,000$ kN/m and is made of steel (A36), and the maximum allowable displacement of the right end of the bar is 0.01 m, determine the minimum allowable cross-sectional area of the bar, A.

PROBLEM 3.4
GIVEN: The uniaxial bar shown below is homogeneous, prismatic and has a distributed axial load of $p_x(x) = p_x^0 x$, where $p_x^0 = $ constant **(note: p_x is not a constant!)** applied along its length as shown.

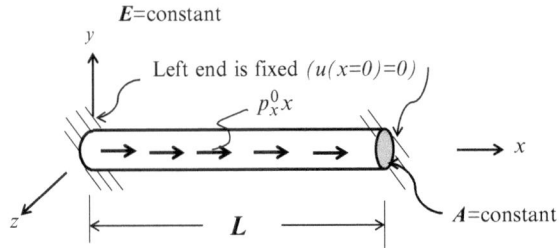

REQUIRED

1. Using uniaxial bar theory, derive an expression for each of the following:

   (a) $P = P(x, p_x^0)$
   (b) $\sigma_{xx} = \sigma_{xx}(x, p_x^0, A, L, E)$
   (c) $\varepsilon_{xx} = \varepsilon_{xx}(x, p_x^0, A, L, E)$
   (d) $u = u(x, p_x^0, A, L, E)$

2. Plot the results of (a)–(d) on four different graphs: $P = P(x)$, $\sigma_{xx} = \sigma_{xx}(x)$, $\varepsilon_{xx} = \varepsilon_{xx}(x)$, and $u = u(x)$ (for a given value of the input loads, geometry, and material properties).

3. Determine the reactions at each end, $P(x = 0)$ and $P(x = L)$.

4. Determine the maximum axial deflection u_{max} and determine its coordinate location.

PROBLEM 3.5

GIVEN: The uniaxial bar shown below is homogeneous, prismatic and has a distributed load $p_x = p_x^0 x$, $p_x^0 = $ constant as shown.

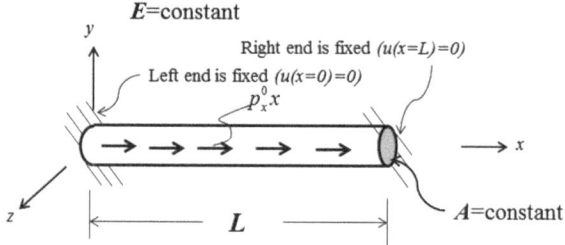

REQUIRED

1. Using uniaxial bar theory, derive an expression for each of the following:

   (a) $u = u(x, p_x^0, A, L, E)$
   (b) $P = P(x, p_x^0, A, L, E)$
   (c) $\sigma_{xx} = \sigma_{xx}(x, p_x^0, A, L, E)$
   (d) $\varepsilon_{xx} = \varepsilon_{xx}(x, p_x^0, A, L, E)$

2. Plot the results of (a)–(d) on four different graphs: $P = P(x)$, $\sigma_{xx} = \sigma_{xx}(x)$, $\varepsilon_{xx} = \varepsilon_{xx}(x)$, and $u = u(x)$ (for a given value of the input loads, geometry, and material properties).

3. Find the location of the maximum stress, σ_{xx} and draw the stress block at that point.

PROBLEM 3.6

GIVEN: The uniaxial bar shown below is homogeneous, prismatic and has a point force, F_x, applied at $x = 2L/3$, as shown.

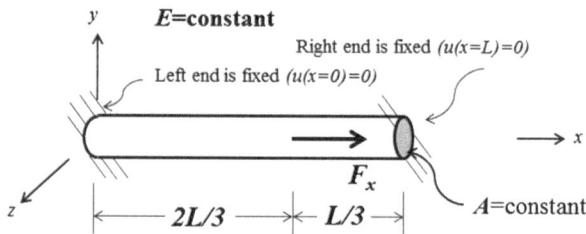

REQUIRED

1. Using uniaxial bar theory, derive an expression for each of the following:

   (a) $u = u(x, F_x, A, L, E)$
   (b) $P = P(x, F_x, A, L, E)$
   (c) $\sigma_{xx} = \sigma_{xx}(x, F_x, A, L, E)$
   (d) $\varepsilon_{xx} = \varepsilon_{xx}(x, F_x, A, L, E)$

2. Plot the results of (a)–(d) on four different graphs: $P = P(x)$, $\sigma_{xx} = \sigma_{xx}(x)$, $\varepsilon_{xx} = \varepsilon_{xx}(x)$, and $u = u(x)$ (for a given value of the input loads, geometry, and material properties).
3. Find the location of the maximum stress, σ_{xx} and draw the stress block at that point.

PROBLEM 3.7
GIVEN: The uniaxial bar shown below is homogeneous, prismatic and has an evenly distributed axial load of $p_x = p_x^0 = $ constant applied along its length, as well as a point force, $F_x = p_x^0 L$, applied at $x = L/2$, as shown.

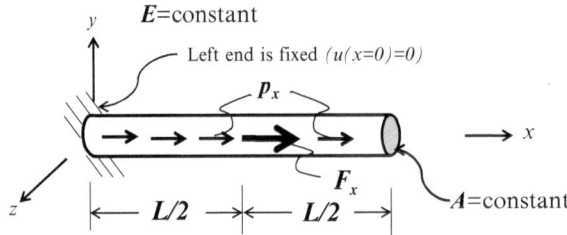

REQUIRED
1. Using uniaxial bar theory, derive an expression for each of the following:

(a) $P = P(x, p_x^0, A, L, E)$
(b) $\sigma_{xx} = \sigma_{xx}(x, p_x^0, A, L, E)$
(c) $\varepsilon_{xx} = \varepsilon_{xx}(x, p_x^0, A, L, E)$
(d) $u = u(x, p_x^0, A, L, E)$

2. Plot the results of (a)–(d) on four different graphs: $P = P(x)$, $\sigma_{xx} = \sigma_{xx}(x)$, $\varepsilon_{xx} = \varepsilon_{xx}(x)$, and $u = u(x)$ (for a given value of the input loads, geometry, and material properties).
3. Find the location of the maximum stress, σ_{xx} and draw the stress block at that point.

PROBLEM 3.8
GIVEN: The uniaxial bar shown below is homogeneous, prismatic and has a distributed axial load of $p_x = p_x^0 x$, $p_x^0 = $ constant applied along its length, as well as a point force, $F_x = p_x^0 L$, applied at $x = L/2$, as shown.

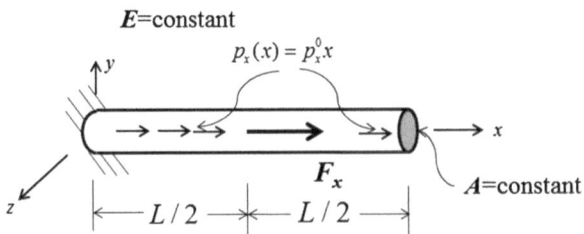

REQUIRED

1. Using uniaxial bar theory, derive an expression for each of the following:

   (a) $P = P(x, p_x^0, A, L, E)$
   (b) $\sigma_{xx} = \sigma_{xx}(x, p_x^0, A, L, E)$
   (c) $\varepsilon_{xx} = \varepsilon_{xx}(x, p_x^0, A, L, E)$
   (d) $u = u(x, p_x^0, A, L, E)$

2. Plot the results of (a)–(d) on four different graphs: $P = P(x)$, $\sigma_{xx} = \sigma_{xx}(x)$, $\varepsilon_{xx} = \varepsilon_{xx}(x)$, and $u = u(x)$ (for a given value of the input loads, geometry, and material properties).

3. Find the location of the maximum stress, σ_{xx}, and draw the stress block at that point.

PROBLEM 3.9

GIVEN: The uniaxial bar shown below is homogeneous, prismatic and has an evenly distributed axial load of $p_x = p_x^0 = $ constant applied along its length, as well as a point force, $F_x = p_x^0 L$, applied at $x = L$.

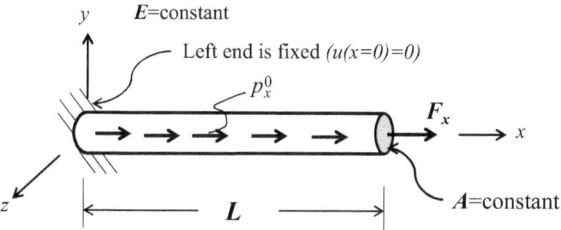

REQUIRED

1. Using uniaxial bar theory, derive an expression for each of the following:

   (a) $P = P(x, p_x^0, A, L, E)$
   (b) $\sigma_{xx} = \sigma_{xx}(x, p_x^0, A, L, E)$
   (c) $\varepsilon_{xx} = \varepsilon_{xx}(x, p_x^0, A, L, E)$
   (d) $u = u(x, p_x^0, A, L, E)$

2. Plot the results of (a)–(d) on four different graphs: $P = P(x)$, $\sigma_{xx} = \sigma_{xx}(x)$, $\varepsilon_{xx} = \varepsilon_{xx}(x)$, and $u = u(x)$ (for a given value of the input loads, geometry, and material properties).

3. Find the location of the maximum stress, σ_{xx} and draw the stress block at that point.

References

Allen D, Haisler W (1985) Introduction to aerospace structural analysis. Wiley, New York

Chapter 4
Theory of Cylindrical Bars Subjected to Torsion

4.1 Introduction

By contrast to the previous chapter on uniaxial bars, this chapter is concerned with the analysis of bars subjected to torsion. While the two subjects are physically different, the reader who has mastered the previous chapter on uniaxial bars will find the current subject mathematically analogous to the previous one.

Recall that we *define a bar as an object that has one dimension that is large compared to the other two.*[1]

If *the bar is subjected uniquely to moments applied about its long dimension, we call the bar a torsion bar*, and in this case we assign the x coordinate axis to be in the direction of the long dimension of the bar, and note that we have used cylindrical coordinates, as shown in Fig. 4.1, due to the geometric shape of the cross-section of the bar. Examples of torsion bars used in structures are shown in Fig. 4.2.

The shape of the cross-sectional area, A, of the bar is not necessarily prismatic, meaning that, in general $A = A(x)$. However, as we will see shortly, *the model to be developed herein will require that the cross-section of the bar be circular*, unlike the model developed previously for uniaxial bars. Furthermore, we will assume that *the bar may be heterogeneous, but the properties do not change in the r or θ coordinate directions.* Thus, in general, $G = G(x)$, $\sigma^T = \sigma^T(x)$, $\sigma^C = \sigma^C(x)$, and $\sigma^S = \sigma^S(x)$, where G is the shear modulus, and σ^T, σ^C, and σ^S are the yield stresses of the material in tension (T), compression (C), and shear (S), as described in Chap. 2. Note also that the geometry of the bar is specified by the length, L, and the cross-sectional area, A. We will see shortly that A will be replaced with a more physically appropriate geometric quantity for torsion called the polar moment of inertia, J. The external loading is described by the axial moment (also called torque) per unit length, m_x, and point torques, T_x, that may be applied at arbitrary coordinate locations along the length of the bar.

---

[1] Note: any italicized statement in the text constitutes a model assumption.

D.H. Allen, *Introduction to the Mechanics of Deformable Solids: Bars and Beams*, DOI 10.1007/978-1-4614-4003-1_4, © Springer Science+Business Media New York 2013

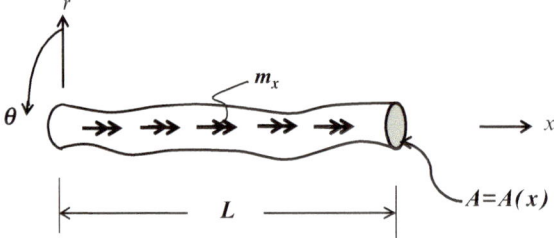

Fig. 4.1 General depiction of a bar subjected torsion loading

Fig. 4.2 Photo of US Army AH-64D Apache Longbow helicopter on *left*; vertical axis wind turbine in Gaspesie, Quebec, Canada on *right*

Table 4.1 Boundary conditions applied to a torsion bar

Boundary conditions
(a)   On the end $x = 0$, either $T_x(x = 0) =$ known or $\theta(x = 0) =$ known
(b)   On the end $x = L$, either $T_x(x = L) =$ known or $\theta(x = L) =$ known

The boundary conditions applied on the ends of the bar may be of two types: either torques, T_x, or circumferential rotations, θ, which can be seen from Fig. 4.1 to be the component of rotation circumferentially about the x-axis. In the current model, it is preferable to use the angle of rotation, θ, instead of the circumferential displacement, u_θ, for reasons that will be explained below. Thus, at each end there is one boundary condition that must be known a priori. These are listed in Table 4.1.

As a review then, all of the inputs required to completely define the problem are described in Table 4.2.

Table 4.2 Torsion bar problem inputs

Problem inputs

1. **Loads**
   (a) $m_x = m_x(x)$, $T_x = T_x(x_1)$
   (b) Boundary conditions: $T_x(x = 0)$ and $T_x(x = L)^{a}$
2. **Geometry**
   (a) $J = J(x)$
   (b) L
   (c) Boundary conditions: $\theta(x = 0)$ and $\theta(x = L)^{a}$
3. **Material properties**
   (a) $G = G(x)$
   (b) $\sigma^T = \sigma^T(x)$, $\sigma^C = \sigma^C(x)$, and $\sigma^S = \sigma^S(x)$

[a]Note that **either** torque **or** displacement is specified at each end

4.2 A Model for Predicting the Mechanical Response of a Cylindrical Torsion Bar

For purposes of creating a robust model that can be utilized to avoid failure of the bar due to fracture or excessive deformation, it is recognized that the following output variables need to be predicted: the stress, $\sigma_{x\theta}$, and the rotation, θ. As we will see, there will be two additional outputs necessary to obtain the above two, and these are: the axial moment, M_x, and the shear strain, $\varepsilon_{x\theta}$. As we will also see, the assumptions that we need in order to construct a simple yet accurate model will require that M_x and θ vary only in the x coordinate direction. Thus, the problem outputs are summarized in Table 4.3.

4.2.1 Construction of the Model

As can be seen from the above listing of outputs, there are four unknowns in the problem. Therefore, **it is evident that we will need four equations in order to construct a rigorous model**. This approach is entirely analogous to the theory of uniaxial bars, and these equations are as follows:

1. Newton's Laws

$$\sum \vec{F} = 0, \quad \sum \vec{M} = 0, \quad \textit{assuming the bar is at rest} \tag{4.1}$$

Note that the above simplifies to the following single equation due to the absence of forces in the x, r, θ and directions, and moments about the r and θ axes.

$$\sum M_x = 0 \tag{4.2}$$

Table 4.3 Torsion bar problem outputs

Problem outputs		
1.	Internal moment:	$M_x = M_x(x)$
2.	Shear stress:	$\sigma_{x\theta} = \sigma_{x\theta}(r, x)$
3.	Shear strain:	$\varepsilon_{x\theta} = \varepsilon_{x\theta}(r, x)$
4.	Circumferential rotation:	$\theta = \theta(x)$

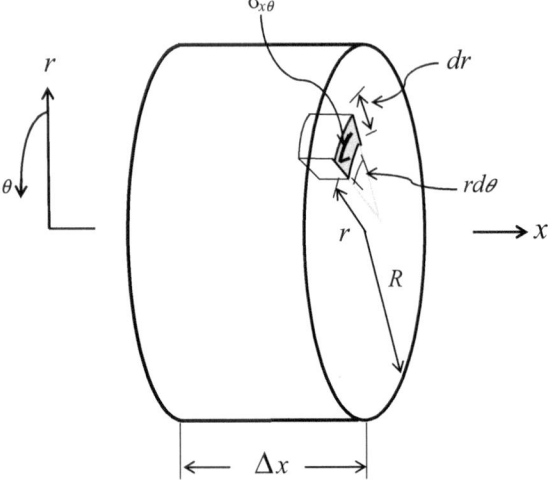

Fig. 4.3 Resultant torque due to shear stress $\sigma_{x\theta}$

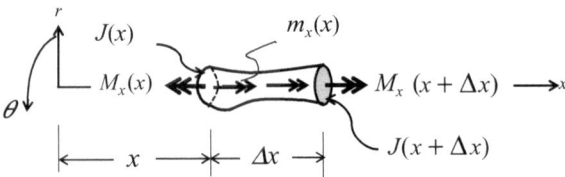

Fig. 4.4 Free body diagram of a section of a torsion bar

Now, let us introduce a new variable, M_x, called the internal axial moment, defined as follows, as depicted in Fig. 4.3.

$$M_x \equiv \int_0^{2\pi} \int_0^R r\sigma_{x\theta} r \, dr \, d\theta \tag{4.3}$$

where R is the radius of the bar, and due to definition (4.3) it is clear that $M = M(x)$.

Next, suppose that a free body diagram of the bar is constructed, depicting two planes passed through the bar normal to the x-axis at coordinate locations x and $x + \Delta x$, as shown in Fig. 4.4.

Equation (4.2) above may now be used to sum moments about the x-axis for the section of the bar shown in Fig. 4.4 as follows:

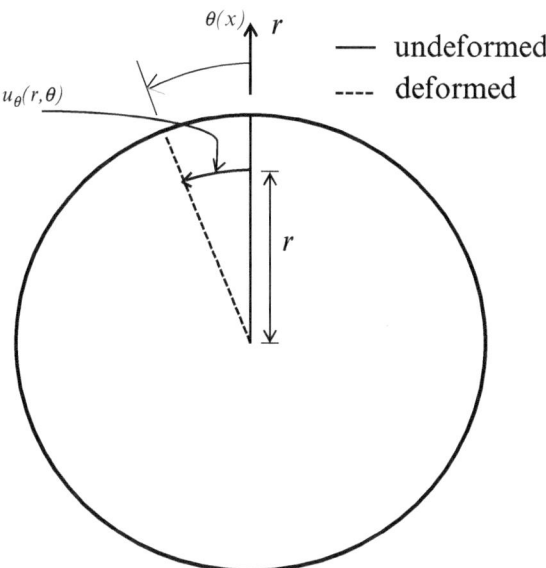

Fig. 4.5 Depiction of rotation of an arbitrary cross-section of a circular torsion bar

$$\sum M_x = 0 \Rightarrow M_x(x + \Delta x) - M_x(x) + \int_x^{x+\Delta x} m_x \, dx = 0 \qquad (4.4)$$

Dividing the above equation through by Δx and invoking the fundamental theorem of calculus, as well as the definition of an ordinary derivative, will result in the following equilibrium equation for the uniaxial bar.

$$\frac{dM_x}{dx} = -m_x(x) \qquad (4.5)$$

2. Kinematics

(a) Strain–displacement relation—

$$\varepsilon_{x\theta} \equiv \frac{du_\theta}{dx} \qquad (4.6)$$

Note that the above is a result of the fact that there is no axial displacement in the bar.

(b) Kinematic assumption—*cross-sections that are planar and normal to the x-axis before loading rotate as a rigid body and remain planar and normal to the x-axis after loading, and the angle of rotation of the cross-section is θ $= \theta(x)$*. The consequence of this assumption is that the circumferential displacement, u_θ, can be expressed in the following separable way, as shown in Fig. 4.5 (Oden and Ripperger 1981):

$$u_\theta(r, \theta) = r\theta \qquad (4.7)$$

Substituting (4.6) into (4.7) now results in the following:

$$\varepsilon_{x\theta} = r\frac{d\theta}{dx} \tag{4.8}$$

It can be seen from the above result that for any given cross-section, the strain is a maximum at the outer surface of the bar.

3. Constitutive Equation

$$\sigma_{x\theta} = G\varepsilon_{x\theta} \Leftrightarrow \varepsilon_{x\theta} = \sigma_{x\theta}/G \tag{4.9}$$

where, as stated above, G is the shear modulus. Equation (4.9) necessarily implies that *the material is orthotropic and must behave linear elastically at all points in the bar*. Note also that as a consequence of our assumptions, all other components of the stress are zero at all points in the bar. Substituting (4.8) into (4.9) results in the following:

$$\sigma_{x\theta} = Gr\frac{d\theta}{dx} \tag{4.10}$$

Equation (4.10) may now be substituted into (4.3), and since $\theta = \theta(x)$, (4.3) will simplify to the following form

$$M_x = GJ\frac{d\theta}{dx} \tag{4.11}$$

where $J = J(x)$, called the polar moment of inertia, is simply a geometric input given by

$$J \equiv \int_0^{2\pi}\int_0^R r^3\, dr\, d\theta \tag{4.12}$$

Note also that (4.10) and (4.11) can be combined to produce the following:

$$\sigma_{x\theta} = \frac{M_x r}{J} \tag{4.13}$$

Now, let us very carefully examine (4.5), (4.8), (4.9), and (4.13). By comparing these equations to information listed in Tables 4.1 and 4.2, it can be seen that there are three kinds of variables listed in these four equations: inputs (*which are known!*), outputs (*which are to be determined!*), and the independent variable, x, which the outputs are to be determined as functions of. For convenience, the four equations are listed in Table 4.4 with the inputs circled and the outputs in boxes.

The above set of equations, together with the boundary conditions described in Table 4.3, constitutes "a well-posed boundary value problem." That is due to the

Table 4.4 Governing equations for torsion bar model

$$\frac{dM_x}{dx} = -\widehat{m_y}(x) \tag{4.5}$$

$$\varepsilon_{x\theta} = r\frac{d\theta}{dx} \tag{4.8}$$

$$\varepsilon_{x\theta} = \sigma_{x\theta}/\widehat{G} \tag{4.9}$$

$$\sigma_{x\theta} = \frac{M_x r}{\widehat{J}} \tag{4.13}$$

fact that there are four equations in four unknowns: M_x, $\sigma_{x\theta}$, $\varepsilon_{x\theta}$, and θ. In addition, there are exactly two derivatives in the equations, thus requiring two boundary conditions, and finally, it can be shown that all four of the equations are mathematically linear, so that one can prove that the above set of equations has a unique solution. *So we have a mathematically acceptable model, but it remains to be seen if it is a physically accurate model.*

For convenience, let us review the assumptions we made in order to construct our model. This is important because if we attempt to use the model to design an object for which any of the assumptions are violated, we are going to necessarily introduce some error into our model. These assumptions, which were previously italicized in the text above, are listed in Table 4.5.

Table 4.5 Assumptions used to construct a model for a circular torsion bar subjected to mechanical loading

Assumptions used to construct the model
1. The object has one dimension that is large compared to the other two (called a bar)
2. The bar is subjected uniquely to moments in the direction of its largest dimension (called a torsion bar)
3. Although not necessarily prismatic, the shape of the cross-section of the bar is circular
4. The material properties do not change in the r or θ coordinate directions (normal to the long axis: x)
5. The bar is at rest
6. Cross-sections that are planar and normal to the x-axis before loading rotate as a rigid body and remain planar and normal to the x-axis after loading
7. The material is orthotropic and must behave linear elastically at all points in the bar

4.2.2 Methods for Obtaining Solutions with the Model

Unfortunately, we are not quite finished yet with our model. We still need to develop systematic methods for solving the four equations shown in Table 4.4 for the four unknowns: M_x, $\sigma_{x\theta}$, $\varepsilon_{x\theta}$, and θ. Before doing that let us make a few observations regarding the equations shown in Table 4.4. These are as follows:

Table 4.6 Systematic method for solving for the unknowns in a statically determinate torsion bar subjected to mechanical loading

Systematic solution method for statically determinate torsion bars
Step 1: Solve (4.5) for $M_x = M_x(x)$ using direct integration (and one torque boundary condition).
Step 2: Use $M_x(x)$ obtained in step 1 to obtain $\sigma_{x\theta} = \sigma_{x\theta}(r, x)$ using (4.13).
Step 3: Use $\sigma_{x\theta}(r, x)$ obtained in step 2 to obtain $\varepsilon_{x\theta} = \varepsilon_{x\theta}(r, x)$ in (4.9).
Step 4: Use $\varepsilon_{x\theta}(r,x)$ obtained in step 3 to obtain $\theta = \theta(x)$ using direct integration in (4.8) (and one rotation boundary condition).

1. Note that the equations are partially coupled, meaning that more than one of the unknowns (in boxes) occurs in several of the equations [except (4.5)]!
2. The equations are all ordinary differential equations in x, but there is one additional independent variable: r.
3. A careful examination of the equations reveals that all of our inputs occur explicitly in the equations in the form of loads: m_x and M_x, geometry: J (and L, via the boundary conditions), and material property: G.
4. As a consequence of observation (3), the resulting four predictive equations for M_x, $\sigma_{x\theta}$, $\varepsilon_{x\theta}$, and θ that we will obtain by solving the four equations in Table 4.4 will turn out to contain the inputs (m_x, J, L, and G) as well as r and x on the right hand side of the four equations. **When we have these equations, we will have a model that we can use to design torsion bars.**

Using the above observations, let us now proceed to construct systematic methods for obtaining equations of the form described in observation (4). As it turns out, there are lots of ways of solving the set of four equations. Sometimes one way is easier than another, and this creates lots of confusion for students. We want to minimize this confusion, so we will develop consistent approaches that will always work.

4.2.2.1 Solution Methods for Statically Determinate Torsion Bars

Recall that either a torque or a rotation boundary condition is applied at each end of the bar but not both. There must be at least one rotation boundary condition in order for the solution to be unique. In the case where a rotation boundary condition is applied at one end of the bar, and a torque boundary condition is applied at the other end, the problem is called "statically determinate." That is due to the fact that in this case one can construct a free body diagram of the entire bar and determine the reaction where the rotation boundary condition is applied by employing equilibrium equation (4.2). The solution method for statically determinate uniaxial bars is described in Table 4.6.

The above method will result in four explicit equations for the unknowns M_x, $\sigma_{x\theta}$, $\varepsilon_{x\theta}$, and θ as functions of the inputs m_x, J, L, and G and the independent variables, r and x. Once these equations are constructed, they can be used to ensure that the bar is designed in such a way that the design constraints (in this course, fracture and excessive rotations) can be satisfied.

There is one remaining question to be answered: is this model accurate? The answer to this question can only be determined by performing systematic experiments that can be used to test the validity of the model. Fortunately, many experiments have been performed to test this theory (beginning in the eighteenth century), and it is generally agreed upon by the scientific and engineering communities that the model is accurate so long as the assumptions listed in Table 4.5 are not violated. Should the designer using this model use the model to design a torsion bar wherein any of the model assumptions are violated, it is recommended that the model be supported by experiments designed to test the accuracy of the model with respect to the assumptions that have been violated in the design process.

Example Problem 4.1
Given: The prismatic and homogeneous torsion bar shown below has a torque T_x applied at the free end ($x = L$).

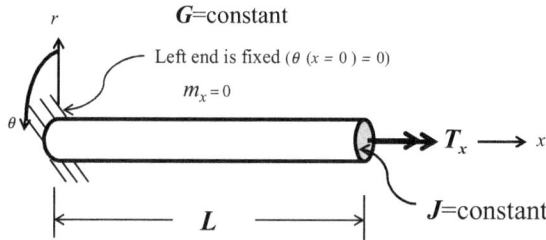

Required

1. Using torsion bar theory, derive an expression for each of the following:

   (a) $M_x = M_x(x, T_x)$
   (b) $\sigma_{x\theta} = \sigma_{x\theta}(r, x, T_x, J, L, G)$
   (c) $\varepsilon_{x\theta} = \varepsilon_{x\theta}(r, x, T_x, J, L, G)$
   (d) $\theta = \theta(x, T_x, J, L, G)$

2. Plot the results of (a)–(d) on four different graphs: M_x vs. x, $\sigma_{x\theta}$ vs. x, $\varepsilon_{x\theta}$ vs. x, and θ vs. x

3. Suppose the bar is 10 m long, with $T_x = 100,000\,\text{kN m}$ and is made of A36 steel ($G = 75\,\text{GPa}$), and the maximum allowable shear stress in the bar is $\sigma_s = 140\,\text{MPa}$, determine the minimum allowable radius, R, of the bar

Solution

1. (a) Since $m_x = 0$, it follows that, from (4.5)

$$\frac{dM_x}{dx} = -m_x = 0 \Rightarrow \int \frac{dM_x}{dx}\,dx = \int 0\,dx \Rightarrow M_x(x) = c_1 \qquad \text{(E.4.1.1)}$$

Now consider the torque, T_x, applied at the right end of the bar. Newton's third law may be used to deduce that the boundary condition at the right end is $M_x(x = L) = T_x$. Applying this boundary condition to (E.4.1.1) results in

$$\boxed{M_x(x) = T_x} \quad \text{Q.E.D.} \qquad \text{(E.4.1.2)}$$

(b) Using (4.13) and (E.4.1.2) results in the following:

$$\sigma_{x\theta} = \frac{M_x r}{J} \Rightarrow \boxed{\sigma_{x\theta} = \frac{T_x r}{J}} \quad \text{Q.E.D.} \tag{E.4.1.3}$$

(c) Using (4.9) and (E.4.1.3) results in the following:

$$\varepsilon_{x\theta} = \frac{\sigma_{x\theta}}{G} \Rightarrow \boxed{\varepsilon_{x\theta} = \frac{T_x r}{GJ}} \quad \text{Q.E.D.} \tag{E.4.1.4}$$

(d) Using (4.8) and (E.4.1.4) results in the following

$$\frac{d\theta}{dx} = \frac{\varepsilon_{x\theta}}{r} = \frac{T_x}{GJ} \Rightarrow \int \frac{d\theta}{dx} dx = \int \frac{T_x}{GJ} dx \Rightarrow \theta(x) = \frac{T_x x}{GJ} + c_2 \tag{E.4.1.5}$$

Applying the boundary condition $\theta(x = 0) = 0$ to (E.4.1.5) implies that $c_2 = 0$. Therefore, (E.4.1.5) simplifies to the following:

$$\boxed{\theta(x) = \frac{T_x x}{GJ}} \quad \text{Q.E.D.} \tag{E.4.1.6}$$

2. Plotting (E.4.1.2)–(E.4.1.4), and (E.4.1.6) results in the following:

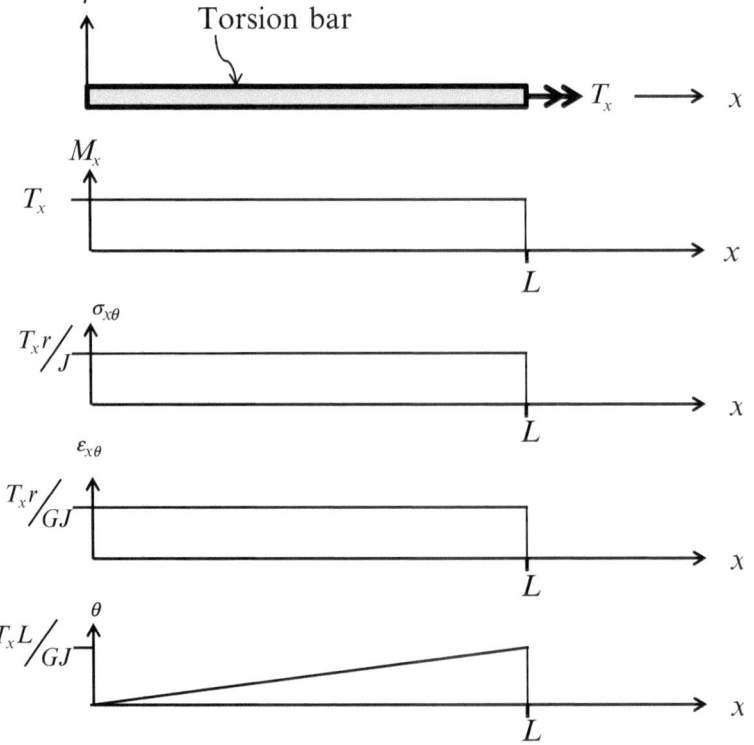

3. The maximum stress occurs at $r = R$ and all values of x, so that (E.4.1.6) gives

$$\sigma_{x\theta_{max}} = \sigma_{x\theta}(r = R) = \frac{T_x R}{J} \qquad \text{(E.4.1.7)}$$

In order to determine J, (4.12) is integrated as follows:

$$J \equiv \int_0^{2\pi} \int_0^R r^3 \, dr \, d\theta = 2\pi \int_0^R r^3 \, dr = \frac{2\pi R^4}{4} = \frac{\pi R^4}{2} \qquad \text{(E.4.1.8)}$$

Substituting (E.4.1.8) into (E.4.1.7) thus results in

$$\sigma_{x\theta_{max}} = \frac{T_x R}{\pi R^4/2} = \frac{2T_x}{\pi R^3} \Rightarrow R_{min} = \sqrt[3]{\frac{2T_x}{\pi \sigma_{x\theta_{max}}}} \qquad \text{(E.4.1.9)}$$

Applying the values from the problem given result in

$$R_{min} = \sqrt[3]{\frac{2 \times 10^8 \text{Nm}}{\pi \times 140 \times 10^6 \text{N/m}^2}} \Rightarrow \boxed{R_{min} = 0.769 \, \text{m}} \quad \text{Q.E.D.} \qquad \text{(E.4.1.10)}$$

4.2.2.2   Solution Methods for Statically Indeterminate Torsion Bars

In the case where rotation boundary conditions are applied at both ends of the torsion bar, the problem is called "statically indeterminate" because there are two reactions and only one equilibrium equation, thus making it intractable to find the reactions using equilibrium equation (4.2) by itself. In this case an alternate solution method is preferable. The method is constructed by substituting (4.11) into (4.5), resulting in the following second-order differential equation in θ.

$$\frac{d}{dx}\left(GJ\frac{d\theta}{dx}\right) = -m_x(x) \qquad (4.14)$$

Since there are two rotation boundary conditions in statically indeterminate problems, (4.14) can be solved directly to obtain the rotation, $\theta = \theta(x)$. The remaining unknowns can be found by back substitution into the other equations, as described in Table 4.7.

Table 4.7 Systematic method for solving for the unknowns in a statically indeterminate torsion bar subjected to mechanical loading

Systematic solution method statically indeterminate torsion bars
Step 1: Solve (4.14) for $\theta = \theta(x)$ using direct integration (and two rotation boundary conditions).
Step 2: Use $\theta(x)$ obtained in step 1 to obtain $M_x = M_x(x)$ using (4.11).
Step 3: Use $M_x = M_x(x)$ obtained in step 2 to obtain $\sigma_{x\theta} = \sigma_{x\theta}(r, x)$ using (4.13).
Step 4: Use $\sigma_{x\theta}(r, x)$ obtained in step 3 to obtain $\varepsilon_{x\theta} = \varepsilon_{x\theta}(r, x)$ using (4.9).

Example Problem 4.2

Given: The prismatic and homogeneous torsion bar shown below has a constant applied moment per unit length $m_x = m_x^0$.

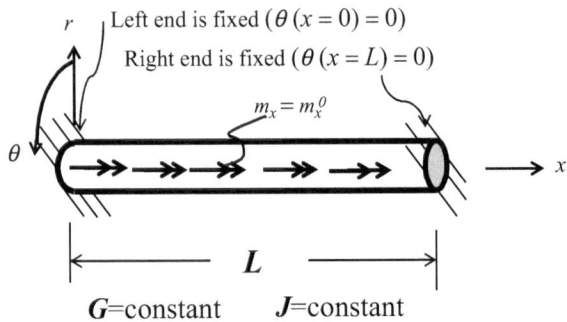

Left end is fixed ($\theta(x = 0) = 0$)

Right end is fixed ($\theta(x = L) = 0$)

$m_x = m_x^0$

L

G=constant J=constant

Required

1. Using torsion bar theory, derive an expression for each of the following:

   (a) $\theta = \theta(x, m_x^0, J, L, G)$
   (b) $M_x = M_x(x, m_x^0)$
   (c) $\sigma_{x\theta} = \sigma_{x\theta}(r, x, m_x^0, J, L, G)$
   (d) $\varepsilon_{x\theta} = \varepsilon_{x\theta}(r, x, m_x^0, J, L, G)$

2. Plot the results of (a)–(d) on four different graphs: M_x vs. x, $\sigma_{x\theta}$ vs. x, $\varepsilon_{x\theta}$ vs. x, and θ vs. x

3. Find the reactions at the left and right ends of the bar

Solution

1. (a) since $m_x^0 = $ constant, it follows that, from (4.16)

$$\frac{d}{dx}\left(GJ\frac{d\theta}{dx}\right) = -m_x^0 \Rightarrow \int \frac{d}{dx}\left(GJ\frac{d\theta}{dx}\right)dx = -\int m_x^0 dx \Rightarrow$$

$$GJ\frac{d\theta}{dx} = -m_x^0 x + C_1 \Rightarrow \int GJ\frac{d\theta}{dx}dx = \int \left(-m_x^0 x + C_1\right)dx \Rightarrow$$

$$\theta(x) = -\frac{m_x^0 x^2}{2GJ} + \frac{C_1 x}{GJ} + C_2 \qquad\qquad\qquad (E.4.2.1)$$

Now consider the boundary condition $\theta(x = 0) = 0$. It can be seen from (E.4.2.1) that

$$C_2 = 0 \tag{E.4.2.2}$$

Next consider the boundary condition $\theta(x = L) = 0$. It can be seen from (E.4.2.1) and (E.4.2.2) that

$$0 = -\frac{m_x^0 L^2}{2GJ} + \frac{C_1 L}{GJ} \Rightarrow C_1 = \frac{m_x^0 L}{2} \tag{E.4.2.3}$$

Substituting (E.4.2.2) and (E.4.2.3) into (E.4.2.1) thus results in

$$\boxed{\theta(x) = -\frac{m_x^0 x^2}{2GJ} + \frac{m_x^0 L x}{2GJ}} \quad \text{Q.E.D.} \tag{E.4.2.4}$$

(b) Substituting (E.4.2.4) into (4.15) results in the following:

$$M_x(x) = GJ\frac{d\theta}{dx} = GJ\frac{d}{dx}\left(-\frac{m_x^0 x^2}{2GJ} + \frac{m_x^0 L x}{2GJ}\right) \Rightarrow$$

$$\boxed{M_x(x) = -m_x^0 x + \frac{m_x^0 L}{2}} \quad \text{Q.E.D.} \tag{E.4.2.5}$$

(c) Substituting (E.4.2.5) into (4.13) results in the following:

$$\boxed{\sigma_{x\theta}(r, x) = \left(-m_x^0 x + \frac{m_x^0 L}{2}\right)\frac{r}{J}} \quad \text{Q.E.D.} \tag{E.4.2.6}$$

(d) Substituting (E.4.2.6) into (4.9) results in the following:

$$\boxed{\varepsilon_{x\theta}(r, x) = \left(-m_x^0 x + \frac{m_x^0 L}{2}\right)\frac{r}{GJ}} \quad \text{Q.E.D.} \tag{E.4.2.7}$$

2. Plotting (E.4.2.4)–(E.4.2.7) results in the following:

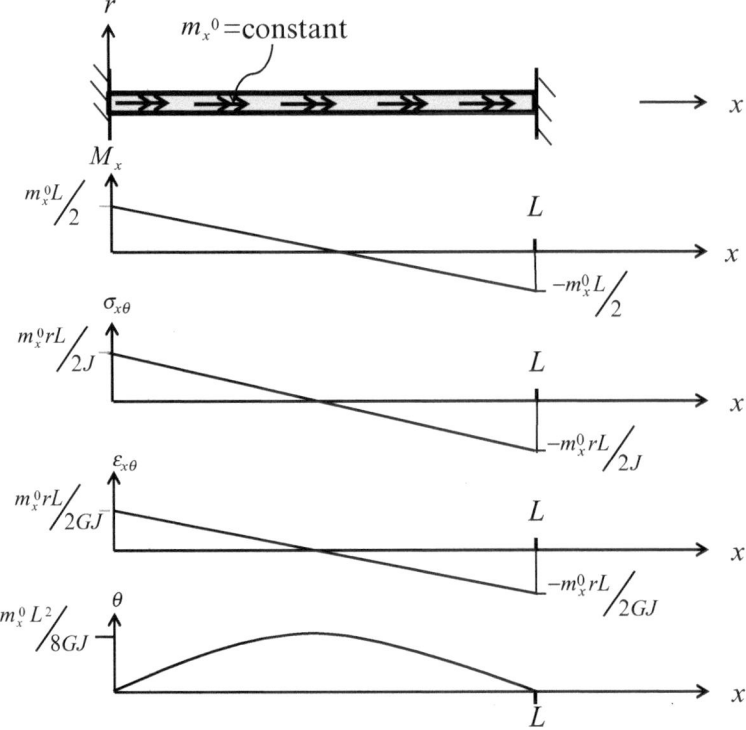

3. It can be seen that the reaction at the left end of the bar, T_x^L, is equal to $M_x(x = 0)$.
   Therefore, using (E.4.2.5) results in

$$T_x^L = M_x(x = 0) = -m_x^0 \times 0 + \frac{m_x^0 L}{2} \Rightarrow \boxed{T_x^L = \frac{m_x^0 L}{2}} \quad \text{Q.E.D.} \qquad \text{(E.4.2.8)}$$

Also, the reaction at the right end of the bar, T_x^R, is equal to $M_x(x = L)$. Therefore,
using (E.4.2.5) again results in

$$T_x^R = M_x(x = L) = -m_x^0 L + \frac{m_x^0 L}{2} \Rightarrow \boxed{T_x^R = -\frac{m_x^0 L}{2}} \quad \text{Q.E.D.} \qquad \text{(E.4.2.9)}$$

4.2.2.3   How to Handle Point Torques Within the Bar

Torsion bars are sometimes subjected to torsion loads that are applied over such
short distances in the x coordinate direction that for practical purposes they can be
considered to be applied at a single point along the x-axis. We term these loads

"point torques." The inclusion of such torques in our model can be accounted for by performing a careful analysis of the kinetics. To see how this can be done, consider a torsion bar with a point torque applied to it, as shown in Fig. 4.6.

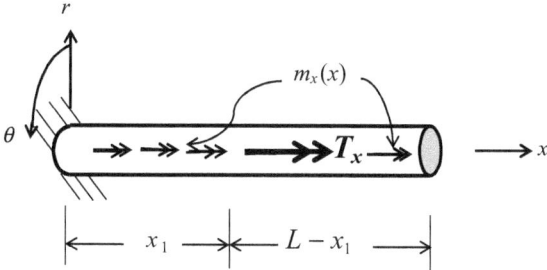

Fig. 4.6 Torsion bar subjected to point torque, T_x

Now, suppose that the bar is cut normal to the x-axis at coordinate location x_1^+, just to the right of the location of the torque, T_x, as shown in Fig. 4.7.

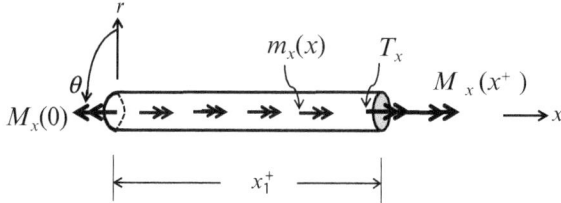

Fig. 4.7 Free body diagram of torsion bar cut to the right of torque, T_x

In this case, summing forces will result in the following:

$$\sum M_x = 0 \Rightarrow M_x(x_1^+) - M_x(0) + T_x + \int_0^{x_1^+} m_x(x)\,dx = 0 \Rightarrow$$

$$M_x(x_1^+) = M_x(0) - T_x - \int_0^{x_1^+} m_x(x)\,dx \qquad (4.15)$$

On the other hand, if the second cut is made just to the left of where the torque, T_x, is applied, the result will be

$$M_x(x_1^-) = M_x(0) - \int_0^{x_1^-} m_x(x)\,dx \qquad (4.16)$$

It follows that since x_1^- and x_1^+ can be arbitrarily close to one another, there must be a jump discontinuity in $M_x(x)$ at the coordinate location x_1. If the torque, T_x, is applied in the positive x direction, then the jump is negative and equal in magnitude to T_x. If the torque, T_x, is applied in the negative x direction, then the jump is positive and equal in magnitude to T_x. The results are summarized in Table 4.8.

Table 4.8 How to handle point torques applied within torsion bars

How to handle a point torque, T_x, applied at coordinate location $x = x_1$
1. If T_x is in the positive x direction, decrease $M_x(x_1)$ by T_x
2. If T_x is in the negative x direction, increase $M_x(x_1)$ by T_x

Example Problem 4.3

Given: The prismatic and homogeneous uniaxial bar shown below has point torques T_x^1 and T_x^2 applied as shown.

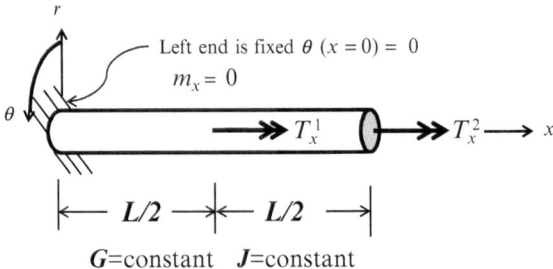

$$G=\text{constant} \quad J=\text{constant}$$

Required

1. Using torsion bar theory, derive an expression for each of the following:

  (a) $M_x = M_x(x, T_x^1, T_x^2, L)$
  (b) $\sigma_{x\theta} = \sigma_{x\theta}(r, x, T_x^1, T_x^2, J, L, G)$
  (c) $\varepsilon_{x\theta} = \varepsilon_{x\theta}(r, x, T_x^1, T_x^2, J, L, G)$
  (d) $\theta = \theta(x, T_x^1, T_x^2, J, L, G)$

2. Plot the results of (a)–(d) on four different graphs: M_x vs. x, $\sigma_{x\theta}$ vs. x, $\varepsilon_{x\theta}$ vs. x, and θ vs. x

Solution

1. Since the problem is statically determinate, use Table 4.6.

  (a) Solve (4.5) for $M_x = M_x(x)$ as follows:

$$\frac{dM_x}{dx} = -m_x = 0 \Rightarrow M_x = C_1 \quad 0 \leq x \leq L/2$$
$$M_x = C_2 \quad L/2 \leq x \leq L \tag{E.4.3.1}$$

  Now apply the boundary condition $M_x(x = L) = T_x^2$. Thus, using Table 4.8, it follows that

$$\boxed{\begin{aligned} M_x &= T_x^1 + T_x^2 \quad 0 \leq x \leq L/2 \\ M_x &= T_x^2 \quad\quad\quad L/2 \leq x \leq L \end{aligned}} \quad \text{Q.E.D.} \tag{E.4.3.2}$$

(b) Solve (4.13) for $\sigma_{x\theta} = \sigma_{x\theta}(r,x)$ as follows

$$\sigma_{x\theta} = \frac{M_x r}{J} \Rightarrow \boxed{\sigma_{x\theta} = \frac{(T_x^1 + T_x^2)r}{J} \quad 0 \le x \le L/2}$$

$$\boxed{\sigma_{x\theta} = \frac{T_x^2 r}{J} \quad L/2 \le x \le L} \quad \text{Q.E.D.} \tag{E.4.3.3}$$

(c) Solve (4.9) for $\varepsilon_{x\theta} = \varepsilon_{x\theta}(r,x)$ as follows:

$$\varepsilon_{x\theta} = \frac{\sigma_{x\theta}}{G} \Rightarrow \boxed{\varepsilon_{x\theta} = \frac{(T_x^1 + T_x^2)r}{GJ} \quad 0 \le x \le L/2}$$

$$\boxed{\varepsilon_{x\theta} = \frac{T_x^2 r}{GJ} \quad L/2 \le x \le L} \quad \text{Q.E.D.} \tag{E.4.3.4}$$

(d) Solve (4.8) for $\theta = \theta(x)$ as follows:

$$\frac{d\theta}{dx} = \frac{\varepsilon_{x\theta}}{r} \Rightarrow \frac{d\theta}{dx} = \frac{(T_x^1 + T_x^2)}{GJ} \quad 0 \le x \le L/2 \Rightarrow$$

$$\theta(x) = \frac{(T_x^1 + T_x^2)x}{GJ} + C_3 \quad 0 \le x \le L/2$$

Now apply the boundary condition $\theta(x = 0) = 0$, from which it is clear that $C_3 = 0$, thus resulting in

$$\boxed{\theta(x) = \frac{(T_x^1 + T_x^2)x}{GJ} \quad 0 \le x \le L/2} \quad \text{Q.E.D.} \tag{E.4.3.5}$$

Next, consider the right half of the bar.

$$\frac{d\theta}{dx} = \frac{T_x^2}{GJ} \Rightarrow \quad \theta(x) = \frac{T_x^2 x}{GJ} + C_4 \quad L/2 \le x \le L \tag{E.4.3.6}$$

Now, it is necessary to match the rotation of the right half of the bar to the left half. To do this, it is necessary to obtain a matching condition for the displacement. This can be obtained by evaluating (E.4.3.5) at $x = L/2$ as follows:

$$\theta(x = L/2) = \frac{(T_x^1 + T_x^2)L}{2EA} \tag{E.4.3.7}$$

The above result may be used as a boundary condition for (E.4.3.6) to obtain C_4 as follows:

$$\theta(x = L/2) = \frac{\left(T_x^1 + T_x^2\right)L}{2GJ} = \frac{T_x^2 L}{2GJ} + C_4 \Rightarrow C_4 = \frac{T_x^1 L}{2GJ} \tag{E.4.3.8}$$

Thus, substituting (E.4.3.8) into (E.4.3.6) results in

$$\boxed{\theta(x) = \frac{T_x^2 x}{GJ} + \frac{T_x^1 L}{2GJ} \quad L/2 \le x \le L} \quad \text{Q.E.D.} \tag{E.4.3.9}$$

2. Plotting (E.4.3.2)–(E.4.3.5), and (E.4.3.9) results in the following graphs.

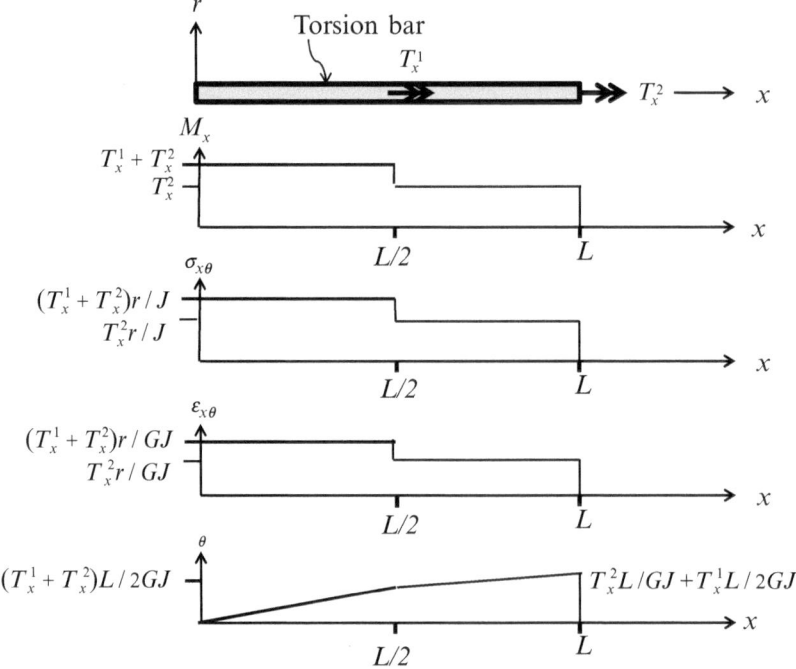

4.3 Assignments

PROBLEM 4.1
GIVEN: The definition of a torsion bar.

REQUIRED: Locate a torsion bar either on the university campus or in the local area, describe it (meaning loads, geometry, and material properties), and include a photo of it.

PROBLEM 4.2
GIVEN: The torsion bar shown below is homogeneous, prismatic and has a point torque, T_x, applied at $x = L/2$ as shown.

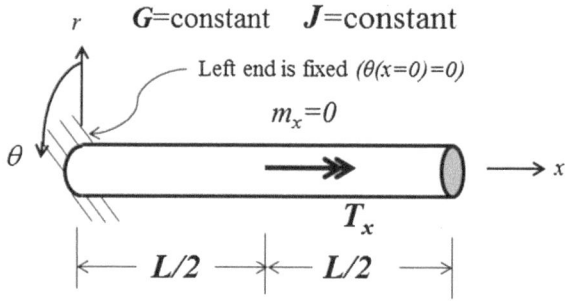

REQUIRED

1. Using torsion bar theory, derive an expression for each of the following:

   (a) $M_x = M_x(x, T_x, J, L, G)$
   (b) $\sigma_{x\theta} = \sigma_{x\theta}(r, x, T_x, J, L, G)$
   (c) $\varepsilon_{x\theta} = \varepsilon_{x\theta}(r, x, T_x, J, L, G)$
   (d) $\theta = \theta(x, T_x, J, L, G)$

2. Plot the results of (a)–(d) on four different graphs: $M_x = M_x(x)$, $\sigma_{x\theta} = \sigma_{x\theta}(x)$, $\varepsilon_{x\theta} = \varepsilon_{x\theta}(x)$, $\theta = \theta(x)$ (for a given value of the input loads, geometry, and material properties and at $r = R$).

3. Determine the location of the maximum shear stress, $\sigma_{x\theta_{max}}$ and draw the stress block with the shear stress denoted on the block.

PROBLEM 4.3

GIVEN: The torsion bar shown below is homogeneous, prismatic and has an evenly distributed torque per unit length of $m_x(x) = m_x^0 = $ constant applied along its length as shown.

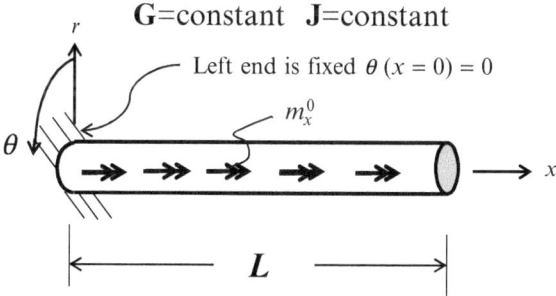

REQUIRED

1. Using torsion bar theory, derive an expression for each of the following:

   (a) $M_x = M_x(x, m_x^0, J, L, G)$
   (b) $\sigma_{x\theta} = \sigma_{x\theta}(x, r, m_x^0, J, L, G)$
   (c) $\varepsilon_{x\theta} = \varepsilon_{x\theta}(x, r, m_x^0, J, L, G)$
   (d) $\theta = \theta(x, m_x^0, J, L, G)$

2. Plot the results of (a)–(d) on four different graphs: $M_x = M_x(x)$, $\sigma_{x\theta} = \sigma_{x\theta}(x)$, $\varepsilon_{x\theta} = \varepsilon_{x\theta}(x)$, $\theta = \theta(x)$ (for a given value of the input loads, geometry, and material properties and at $r = R$).
3. Determine the location of the maximum shear stress, $\sigma_{x\theta_{max}}$ and draw the stress block with the shear stress denoted on the block.

PROBLEM 4.4

GIVEN: The torsion bar shown below is homogeneous, prismatic and has a distributed torque per unit length of $m_x(x) = m_x^0 x$, where $m_x^0 = $ constant **(note: m_x is not a constant!)** applied along its length as shown.

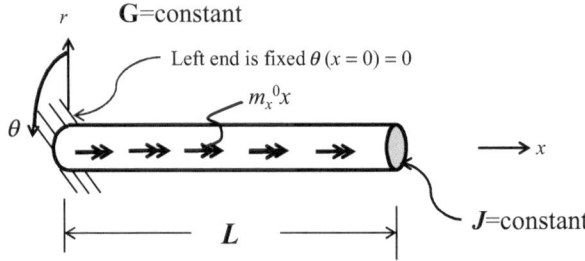

REQUIRED

1. Using torsion bar theory, derive an expression for each of the following:

   (a) $M_x = M_x(x, m_x^0)$
   (b) $\sigma_{x\theta} = \sigma_{x\theta}(r, x, m_x^0, J, L, G)$
   (c) $\varepsilon_{x\theta} = \varepsilon_{x\theta}(r, x, m_x^0, J, L, G)$
   (d) $\theta = \theta(x, m_x^0, J, L, G)$

2. Plot the results of (a)–(d) on four different graphs: $M_x = M_x(x)$, $\sigma_{x\theta} = \sigma_{x\theta}(x)$, $\varepsilon_{x\theta} = \varepsilon_{x\theta}(x)$, $\theta = \theta(x)$ (for a given value of the input loads, geometry, and material properties).
3. Determine the maximum axial deflection θ_{max} and its coordinate location.

PROBLEM 4.5

GIVEN: The torsion bar shown below is homogeneous, prismatic and has a distributed torque per unit length of $m_x(x) = m_x^0 x$, where $m_x^0 = $ constant **(note: m_x is not a constant!)** applied along its length as shown.

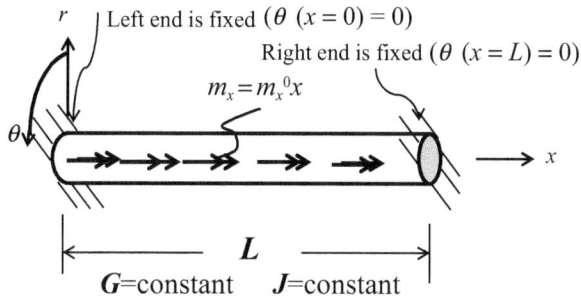

REQUIRED

1. Using torsion bar theory, derive an expression for each of the following:

   (a) $\theta = \theta(x,\ m_x^0,\ J,\ L,\ G)$
   (b) $M_x = M_x(x,\ m_x^0,\ J,\ L,\ G)$
   (c) $\sigma_{x\theta} = \sigma_{x\theta}(r,\ x,\ m_x^0,\ J,\ L,\ G)$
   (d) $\varepsilon_{x\theta} = \varepsilon_{x\theta}(r,\ x,\ m_x^0,\ J,\ L,\ G)$

2. Plot the results of (a)–(d) on four different graphs: $M_x = M_x(x)$, $\sigma_{x\theta} = \sigma_{x\theta}(x)$, $\varepsilon_{x\theta} = \varepsilon_{x\theta}(x)$, $\theta = \theta(x)$ (for a given value of the input loads, geometry, and material properties).

3. Find the reactions at each end of the bar.

PROBLEM 4.6

GIVEN: The torsion bar shown below is homogeneous, prismatic and has a point torque, T_x, applied at $x = 2L/3$, as shown.

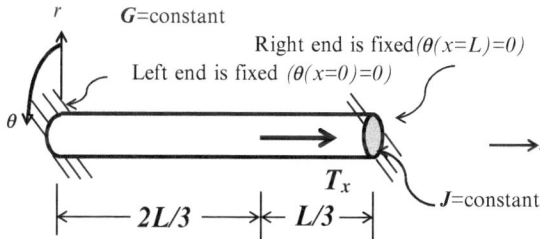

REQUIRED

1. Using torsion bar theory, derive an expression for each of the following:

   (a) $\theta = \theta(x,\ T_x,\ J,\ L,\ G)$
   (b) $M_x = M_x(x,\ T_x,\ J,\ L,\ G)$
   (c) $\sigma_{x\theta} = \sigma_{x\theta}(x,\ r,\ T_x,\ J,\ L,\ G)$
   (d) $\varepsilon_{x\theta} = \varepsilon_{x\theta}(x,\ r,\ T_x,\ J,\ L,\ G)$

2. Plot the results of (a)–(d) on four different graphs: $M_x = M_x(x)$, $\sigma_{x\theta} = \sigma_{x\theta}(x)$, $\varepsilon_{x\theta} = \varepsilon_{x\theta}(x)$, and $\theta = \theta(x)$ (for a given value of the input loads, geometry, and material properties).
3. Find the location of the maximum stress, $\sigma_{x\theta}$ and draw the stress block at that point.

PROBLEM 4.7
GIVEN: The torsion bar shown below is homogeneous, prismatic and has an evenly distributed moment per unit length of $m_x = m_x^0 = $ constant applied along its length, as well as a point torque, $T_x = m_x^0 L$, applied at $x = L/2$, as shown.

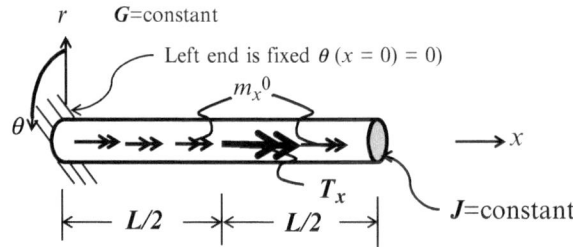

REQUIRED

1. Using torsion bar theory, derive an expression for each of the following:

   (a) $M_x = M_x(x, m_x^0, J, L, G)$
   (b) $\sigma_{x\theta} = \sigma_{x\theta}(x, r, m_x^0, J, L, G)$
   (c) $\varepsilon_{x\theta} = \varepsilon_{x\theta}(x, r, m_x^0, J, L, G)$
   (d) $\theta = \theta(x, m_x^0, J, L, G)$

2. Plot the results of (a)–(d) on four different graphs: $M_x = M_x(x)$, $\sigma_{x\theta} = \sigma_{x\theta}(x)$, $\varepsilon_{x\theta} = \varepsilon_{x\theta}(x)$, and $\theta = \theta(x)$ (for a given value of the input loads, geometry, and material properties).
3. Find the location of the maximum stress, $\sigma_{x\theta}$ and draw the stress block at that point.

PROBLEM 4.8
GIVEN: The torsion bar shown below is homogeneous, prismatic and has a distributed axial moment per unit length of $m_x = m_x^0 x$, $m_x^0 = $ constant applied along its length, as well as a point torque, $T_x = m_x^0 L$, applied at $x = L/2$, as shown.

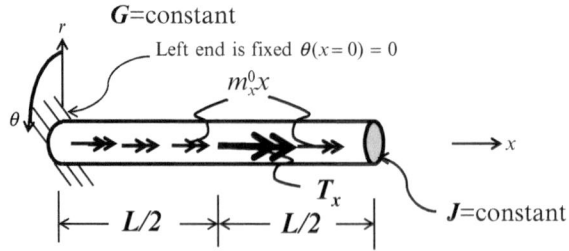

REQUIRED

1. Using torsion bar theory, derive an expression for each of the following:

   (a) $M_x = M_x(x, m_x^0, J, L, G)$
   (b) $\sigma_{x\theta} = \sigma_{x\theta}(x, r, m_x^0, J, L, G)$
   (c) $\varepsilon_{x\theta} = \varepsilon_{x\theta}(x, r, m_x^0, J, L, G)$
   (d) $\theta = \theta(x, r, m_x^0, J, L, G)$

2. Plot the results of (a)–(d) on four different graphs: $M_x = M_x(x)$, $\sigma_{x\theta} = \sigma_{x\theta}(x)$, $\varepsilon_{x\theta} = \varepsilon_{x\theta}(x)$, and $\theta = \theta(x)$ (for a given value of the input loads, geometry, and material properties).

3. Find the location of the maximum stress, $\sigma_{x\theta}$ and draw the stress block at that point.

PROBLEM 4.9

GIVEN: The torsion bar shown below is homogeneous, prismatic and has an evenly distributed axial moment per unit length of $m_x = m_x^0 = $ constant applied along its length, as well as a point torque, $T_x = m_x^0 L$, applied at $x = L$, as shown.

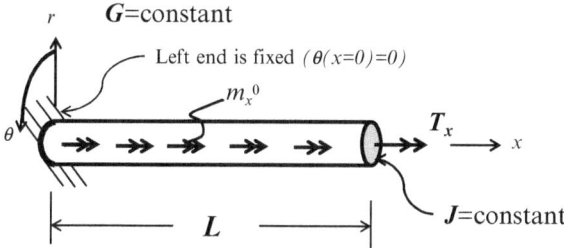

REQUIRED

1. Using torsion bar theory, derive an expression for each of the following:

   (a) $M_x = M_x(x, m_x^0, J, L, G)$
   (b) $\sigma_{x\theta} = \sigma_{x\theta}(x, r, m_x^0, J, L, G)$
   (c) $\varepsilon_{x\theta} = \varepsilon_{x\theta}(x, r, m_x^0, J, L, G)$
   (d) $\theta = \theta(x, m_x^0, J, L, G)$

2. Plot the results of (a)–(d) on four different graphs: $M_x = M_x(x)$, $\sigma_{x\theta} = \sigma_{x\theta}(x)$, $\varepsilon_{x\theta} = \varepsilon_{x\theta}(x)$, and $\theta = \theta(x)$ (for a given value of the input loads, geometry, and material properties).

3. Find the location of the maximum stress, $\sigma_{x\theta}$ and draw the stress block at that point.

References

Oden J, Ripperger E (1981) Mechanics of elastic structures, 2nd edn. McGraw-Hill, New York

Chapter 5
Theory of Beams

5.1 Introduction

The reader will recall that in the two previous chapters on uniaxial bars and torsion bars there were significant mathematical similarities despite the fact that the physics of those two models are markedly different. Such mathematical similarities occur often in nature for problems that bear little physical resemblance to one another. In this chapter, we consider the theory of beams, and while this theory does bear some mathematical similarity to the models developed in the two previous chapters, the theory of beams is significantly more complicated than what we have heretofore studied in this text.

Recall that we *define a bar as an object that has one dimension that is large compared to the other two.*[1] If *the bar is uniquely loaded normal to the direction of its long dimension*, we call it a beam, and as in previous cases we assign the x coordinate axis to be in the direction of the long dimension of the bar, as shown in Fig. 5.1. Correspondingly, in this course the y coordinate axis is assigned to be collinear with the direction of the externally applied forces.

It may be argued that beams represent the single most commonly employed structural component in our world today, as evidenced by the examples shown in Fig. 5.2. Therefore, it is essential that the student who aspires to be a structural engineer develop a concise understanding of the materials contained in this chapter.

Note that the shape of the cross-sectional area, A, of the beam shown in Fig. 5.1 is not necessarily circular, nor is the beam prismatic, meaning that, in general $A = A(x)$. Furthermore, we will assume that *the beam may be heterogeneous, but the properties do not change in the y or z coordinate directions.* Thus, in general, $E = E(x)$, $\sigma^T = \sigma^T(x)$, $\sigma^C = \sigma^C(x)$, and $\sigma^S = \sigma^S(x)$, where E is Young's modulus, and σ^T, σ^C, and σ^S are the yield stresses of the material in tension (T), compression (C), and shear (S), as described in Chap. 2. Note also that the geometry of the beam

---

[1] Note: any italicized statement in the text constitutes a model assumption.

D.H. Allen, *Introduction to the Mechanics of Deformable Solids: Bars and Beams*,
DOI 10.1007/978-1-4614-4003-1_5, © Springer Science+Business Media New York 2013

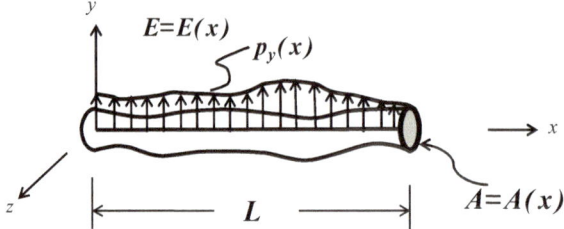

Fig. 5.1 General depiction of a beam subjected to mechanical loading

Fig. 5.2 Modern structures constructed with beams: The Cathedral of Brasilia by Oscar Niemeyer; The Eiffel Tower by Gustav Eiffel

is specified by the length, L, and the cross-sectional area, A. As we will see shortly, the geometric property that will be of significance for the theory of beams is the moment of inertia, $I_{zz} = I_{zz}(x)$. The external loading is described by the transverse loading per unit length, $p_y = p_y(x)$ (which may include gravitational loads), and transverse point loads, F_y, that may be applied at the ends of the bar ($x = 0$, L) as well as at other points along the beam. There may also be moments applied about the z-axis, denoted M_z.

Note that when a bar is subjected to transverse loading such as that shown in Fig. 5.1, the bar will necessarily respond to that load by bending along its length. Thus, it should be apparent that for loads applied in the x-y plane, there will be a component of deformation of the beam, $v = v(x,\ y,\ z)$, in the y coordinate direction, and this component is called the transverse deflection. In this chapter, we will concern ourselves with the vertical deflection of the centroidal axis, termed $v_0 \equiv v$ (x, $y = 0$, $z = 0$) $= v_0(x)$ for reasons that will become clear below. Furthermore, the beam will become curved due to the loading, so that the long dimension of the beam will not remain parallel to the x-axis. In other words, the beam will be sloped, as expressed by $dv_0/dx = dv_0/dx(x)$.

As we have discussed in previous chapters, boundary conditions applied at the ends of a bar may be of two types: either kinetic or kinematic. In the case of a beam, it is physically possible to almost perfectly restrain the ends of the beam against both displacement and rotation. Nonetheless, the reader is reminded that the reality of what happens physically at the boundary is rarely exactly as it is depicted in a mathematical model intended to simulate reality.

According to the model to be developed herein, at a given boundary of the beam there may be two kinematic boundary conditions, unlike a uniaxial bar, which only has one kinematic displacement boundary condition at a given boundary. Furthermore, it is also possible to apply two kinetic boundary conditions at a given boundary of a beam: force in the y coordinate direction and moment about the z-axis. However, due to the physical nature of the problem, only two boundary conditions are possible at any boundary. These two may be purely kinematic (called essential boundary conditions), purely kinetic (called natural boundary conditions), or one kinematic and one kinetic (called mixed). These types of boundary conditions are usually depicted by simple cartoons, as shown in Fig. 5.3.

Examples of simply supported (pinned) and fixed boundary conditions applied to beams are shown in Figs. 5.4 and 5.5, respectively.

The possible boundary conditions required within the model are listed in Table 5.1.

As a review then, all of the inputs required to completely define the problem are described in Table 5.2.

5.2 A Model for Predicting the Mechanical Response of a Beam

When a beam is subjected to transverse loading in the y coordinate direction, there are two components of stress that are known from experimental observation to be significant, and these are σ_{xx} and σ_{xy}, as shown in Fig. 5.6.

Now, suppose that we define the following kinetic variables in accordance with Fig. 5.6.

$$V_y \equiv \int\int \sigma_{xy}\,dy\,dz \tag{5.1}$$

where V_y is called the internal shear resultant, and it is clear from (5.1) that $V_y = V_y(x)$. Furthermore, define

$$M_z \equiv -\int\int y\sigma_{xx}\,dy\,dz \tag{5.2}$$

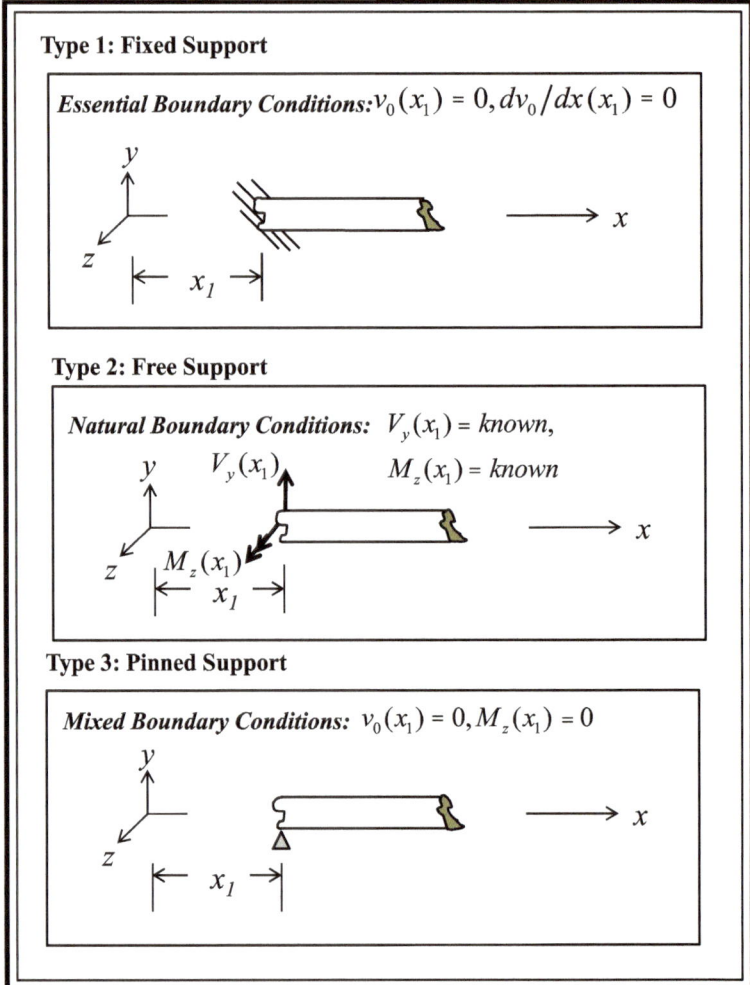

Fig. 5.3 Cartoons depicting different beam boundary conditions

where it can be seen that we have introduced the negative sign in (5.2) in order to be consistent with the right hand rule. M_z is called the internal resultant bending moment, and it is clear from (5.2) that $M_z = M_z(x)$.

For purposes of creating a robust model that can be utilized to avoid failure of the beam due to fracture or excessive deformation, the following output variables will need to be predicted: the internal resultants V_y and M_z; the stress components σ_{xx} and σ_{xy}; the strain components ε_{xx} and ε_{xy}; and the displacement v_0 and slope dv_0/dx. Thus, the problem outputs are summarized in Table 5.3.

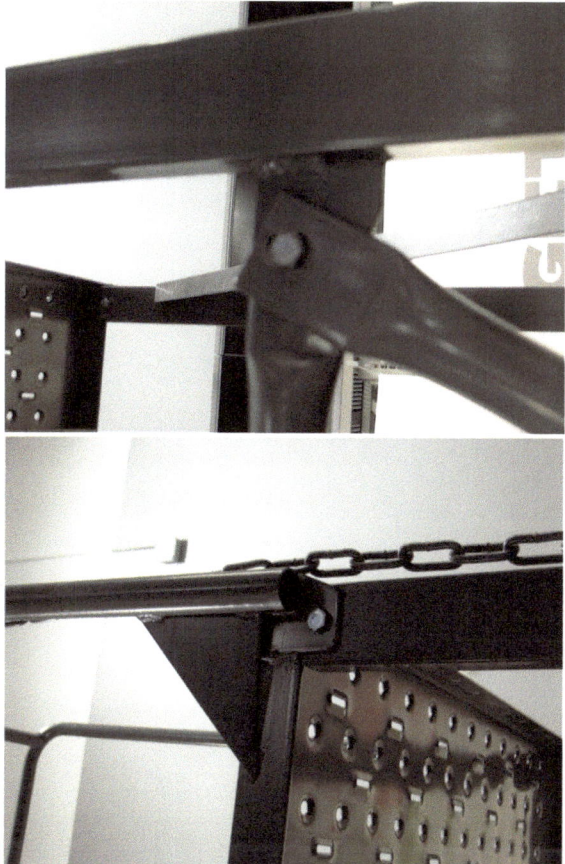

Fig. 5.4 Examples of pinned beam connections

5.2.1   Construction of the Model

As can be seen from the above listing of outputs, there are eight unknowns in the problem. Therefore, **it is evident that we will need eight equations in order to construct a rigorous model**. In addition to (5.1) and (5.2), these are as follows.

(1) **Newton's Laws**

$$\sum \vec{F} = 0, \ \sum \vec{M} = 0, \quad \textit{assuming the bar is at rest} \tag{5.3}$$

Note that the above simplifies to the following equations due to the absence of forces in the *x* and *z* directions and moments about the *x* and *y* axes.

$$\sum F_y = 0, \ \sum M_y = 0 \tag{5.4}$$

Fig. 5.5 Examples of fixed beam connections. *Clockwise from top left*: Connection for a large (30 m tall) highway high mast light showing 1 m diameter base; Welded connections on a portable ladder; Bronze statue of a colt showing front right leg used as a cantilever (from the bronze sculpture entitled "Mustangs at Las Colinas" by artist Robert Glen located at Williams Square in Irving, Texas); and Field of large concrete cantilevered beam–columns at a highway interchange

Table 5.1 Boundary conditions applied to a beam

Boundary conditions
(a) On the end $x = 0$, either $V_y(x = 0) = $ known, $M_z(x = 0) = $ known or $v_0(x = 0) = $ known, $dv_0/dx(x = 0) = $ known
(b) On the end $x = L$, either $V_y(x = L) = $ known, $M_z(x = L) = $ known or $v_0(x = L) = $ known, $dv_0/dx(x = L) = $ known

Next, suppose that a free body diagram of the beam is constructed, depicting two planes passed through the bar normal to the x-axis at coordinate locations x and $x + \Delta x$, as shown in Fig. 5.7.

Equation (5.4) may now be used to sum forces in the y direction for the section of the bar shown in Fig. 5.10 as follows:

Table 5.2 Beam problem inputs

Problem inputs

1. **Loads:**
   (a) $p_y = p_y(x), F_y(x_1),$ *and* $M_z(x_2)$
   (b) Boundary conditions: $V_y(x = 0), M_z(x = 0), V_y(x = L)$ and $M_z(x = L)^a$
2. **Geometry:**
   (a) $I_{zz} = I_{zz}(x)$
   (b) L
   (c) Boundary conditions $v_0(x = 0), dv_0/dx(x = 0), v_0(x = L),$ and $dv_0/dx(x = L)^a$
3. **Material properties**
   (a) $E = E(x)$
   (b) $\sigma^T = \sigma^T(x), \sigma^C = \sigma^C(x), \sigma^S = \sigma^S(x)$

^{a}Note that **only two boundary conditions** are specified at each end

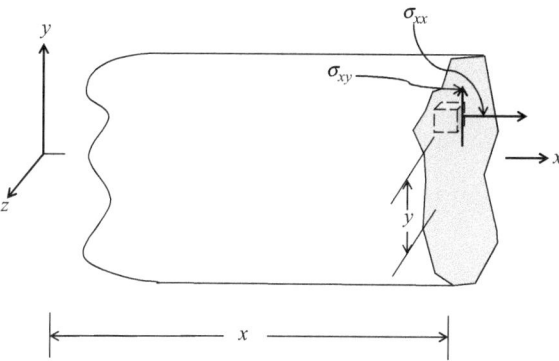

Fig. 5.6 Cross-section of a beam showing stress state

Table 5.3 Beam problem outputs

Problem outputs

1. Internal resultants: $V_y = V_y(x)$ and $M_z = M_z(x)$
2. Stress components: $\sigma_{xx} = \sigma_{xx}(x, y)$ and $\sigma_{xy} = \sigma_{xy}(x, y)$
3. Strain components: $\varepsilon_{xx} = \varepsilon_{xx}(x, y)$ and $\varepsilon_{xy} = \varepsilon_{xy}(x, y)$
4. Displacement and slope: $v_0 = v_0(x)$ and $dv_0/dx = dv_0/dx(x)$

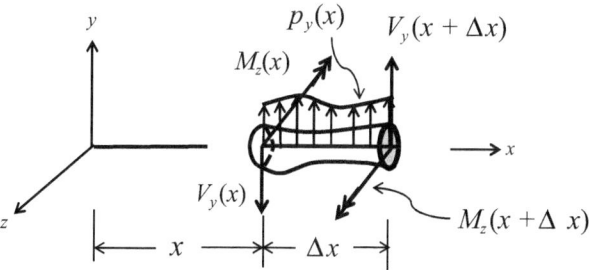

Fig. 5.7 Free body diagram of a section of a beam

$$\sum F_y = 0 \Rightarrow V_y(x + \Delta x) - V_y(x) + \int_x^{x+\Delta x} p_y \, dx = 0 \qquad (5.5)$$

Dividing the above equation through by Δx and invoking the fundamental theorem of calculus as well as the definition of an ordinary derivative will result in the following equilibrium equation for the uniaxial bar.

$$\frac{dV_y}{dx} = -p_y(x) \qquad (5.6)$$

Similarly, summing moments about the z-axis results in the following:

$$M_z(x + \Delta x) - M_z(x) + \Delta x V_y(x) - \int_x^{x+\Delta x} \alpha \Delta x p_y(x) \, dx = 0 \qquad (5.7)$$

where $\alpha \Delta x$ is the moment arm for the distributed load $p_y(x)$ and $0 \le \alpha \le 1$. Dividing the above equation through by Δx and taking the limit as x approaches zero will result in the following equilibrium equation.

$$\frac{dM_z}{dx} = -V_y \qquad (5.8)$$

where the last term vanishes due to the inclusion of Δx inside the integral.

(2) **Kinematics**

(a) Strain–displacement relations—

$$\varepsilon_{xx} \equiv \frac{\partial u}{\partial x} \qquad (5.9)$$

and

$$\varepsilon_{xy} \equiv \frac{\partial u}{\partial y} + \frac{\partial v}{\partial x} \qquad (5.10)$$

(b) Kinematic assumption—*cross-sections that are planar and normal to the x-axis before loading remain planar and normal to the centroidal axis of the beam after loading.* This assumption, while it is nearly identical to the assumption made in the previous two chapters, is immediately seen to be much more profound for a beam because the beam is seen to deform into the shape of a curve, so that the plane sections are not parallel to one another in the deformed beam, as shown in Fig. 5.8. The above assumption is known as the Euler–Bernoulli assumption, because history records that it was first proposed by Daniel Bernoulli to Leonhard Euler in a letter written circa

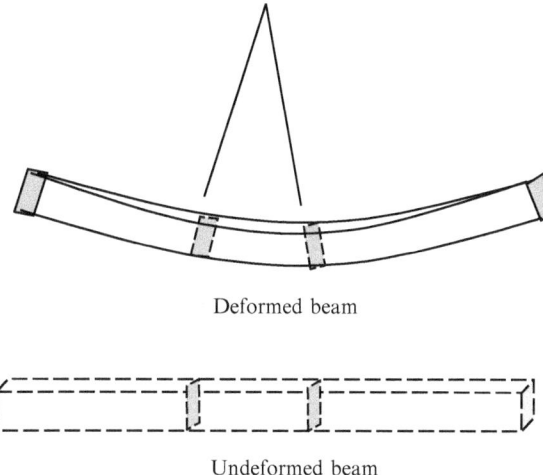

Fig. 5.8 A beam deforming according to the Euler–Bernoulli assumption

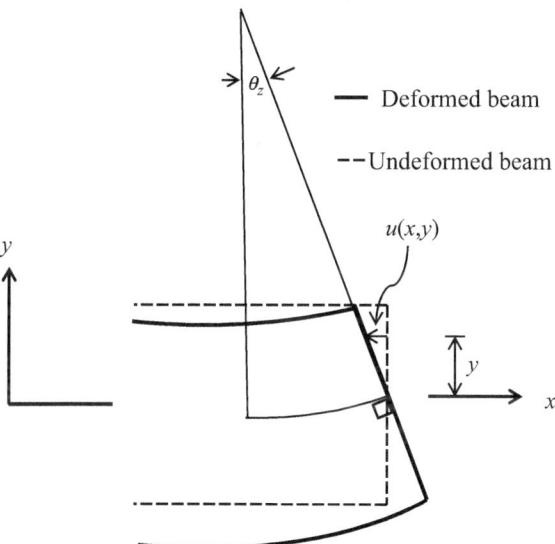

Fig. 5.9 View of Euler–Bernoulli assumption showing displacement, $u(x, y)$

1740 (Euler 1744). It is one of the first great simplifying assumptions made in the field of mechanics. The consequence of this assumption is that the axial displacement, u, may be described by the following equation, as depicted in Fig. 5.9.

$$u(x, y) = -\theta_z(x)y \tag{5.11}$$

where $\theta_z(x)$ is the angle of rotation of the initially vertical planes about the z-axis, as depicted in the figure.

(3) Constitutive equations

$$\sigma_{xx} = E\varepsilon_{xx} \Leftrightarrow \varepsilon_{xx} = \sigma_{xx}/E \tag{5.12}$$

$$\sigma_{xy} = G\varepsilon_{xy} \Leftrightarrow \varepsilon_{xy} = \sigma_{xy}/G \tag{5.13}$$

which necessarily implies that *the material is orthotropic and must behave linear elastically at all points in the bar* and that $\sigma_{xx} \gg \sigma_{yy}, \sigma_{zz}$.

5.2.2   The Beam Model for the Transverse Deflection

Now note that, as a necessary consequence of assumption (5.11) above, (5.9) may be written as follows:

$$\varepsilon_{xx} = -y\frac{d\theta_z}{dx} \tag{5.14}$$

Substituting (5.14) into (5.12) thus results in the following:

$$\sigma_{xx} = -Ey\frac{d\theta_z}{dx} \tag{5.15}$$

The above equation may now be substituted into (5.2), resulting in

$$M_z = -\int\int\left(-Ey\frac{d\theta_z}{dx}\right)y\,dy\,dz = \int\int E\frac{d\theta_z}{dx}y^2\,dy\,dz \tag{5.16}$$

It has been previously assumed that the material is heterogeneous only in the x coordinate direction. Using this assumption, (5.16) may be written as follows:

$$M_z = EI_{zz}\frac{d\theta_z}{dx} \tag{5.17}$$

where, I_{zz}, called the second area moment of inertia about the z axis, is given by

$$I_{zz} \equiv \int\int y^2\,dy\,dz = \int y^2\,dA \tag{5.18}$$

Substituting (5.15) into (5.17) will result in

$$\sigma_{xx} = \frac{-M_z y}{I_{zz}} \tag{5.19}$$

In order to calculate the vertical component of displacement recall from calculus that *for small displacements in the y coordinate direction* (Greenberg 1978; Allen and Haisler 1985)

$$\frac{d^2 v_0}{dx^2} \cong \frac{d\theta_z}{dx} \tag{5.20}$$

Therefore, substituting (5.20) into (5.17) will result in

$$\frac{d^2 v_0}{dx^2} = \frac{M_z}{EI_{zz}} \tag{5.21}$$

From (5.19)–(5.21), it is possible to gain some insight into the relation between the kinetics (meaning the moment, $M_z(x)$, and the stress, $\sigma_{xx}(x,y)$) and the kinematics (meaning the displacement, $v_0(x)$, and the curvature, $d^2 v_0/dx^2(x)$). The observant student of the subject will gain some preconceived notion as to the deformations in a given beam due to the externally applied loads, and from this it is possible to guess the sign, and to a certain extent, the magnitude of the moment and stress in the beam, as shown for an example problem in Fig. 5.10.

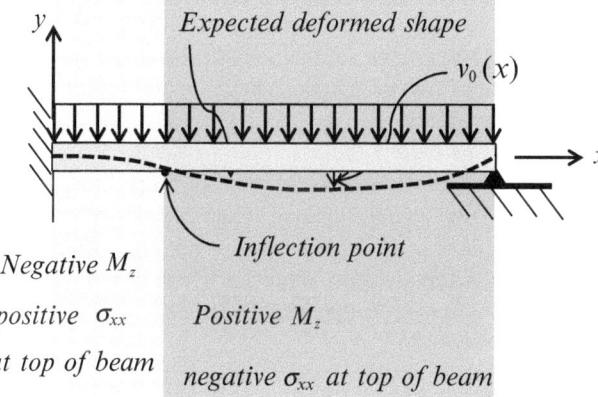

Fig. 5.10 Using the deformation inflection point to guess the sign of the bending moment and stress in a beam

Using the results obtained thus far, it is now possible to construct a major portion of the model for a beam. This can be accomplished by carefully examining (5.6), (5.8), (5.12), (5.19), and (5.21). By comparing these equations to information listed in Tables 5.2 and 5.3, it can be seen that there are three kinds of variables listed in these five equations: inputs (*which are known!*), outputs (*which are to be determined!*), and the independent variables, x and y, which the outputs are to be determined as functions of. For convenience, the five equations are listed in Table 5.4 with the inputs circled and the outputs in boxes.

Table 5.4 Governing equations for beam transverse deflection model

$$\frac{d\boxed{V_y}}{dx} = \boxed{-p_y}(x) \tag{5.6}$$

$$\frac{d\boxed{M_z}}{dx} = \boxed{-V_y} \tag{5.8}$$

$$\boxed{\varepsilon_{xx}} = \frac{\boxed{\sigma_{xx}}}{\boxed{E}} \tag{5.12}$$

$$\boxed{\sigma_{xx}} = \frac{\boxed{M_z}y}{\boxed{I_{zz}}} \tag{5.19}$$

$$\frac{d^2\boxed{v_0}}{d_x{}^2} = \frac{\boxed{M_z}}{\boxed{EI_{zz}}} \tag{5.21}$$

The above set of equations, together with the boundary conditions described in Table 5.1, constitutes a well-posed boundary value problem. That is due to the fact that there are five equations in five unknowns: $V_y = V_y(x)$, $M_z = M_z(x)$, $\sigma_{xx} = \sigma_{xx}(x, y)$, $\varepsilon_{xx} = \varepsilon_{xx}(x, y)$, and $v_0 = v_0(x)$. In addition, there are exactly four derivatives in the equations, thus requiring four boundary conditions, and finally, it can be shown that all five of the equations are mathematically linear (see Chap. 7), so that one can prove that the above set of equations has a unique solution. Thus, we have a mathematically acceptable model.

For convenience, let us review the assumptions we made in order to construct our model. This is important because if we attempt to use the model to design an object for which any of the assumptions are violated, we are going to necessarily introduce some error into our model. These assumptions, which were previously italicized in the text above, are listed in Table 5.5. There is an extensive literature on the theories of beams (Oden and Ripperger 1981; Popov 1998; Wempner 1995).

Table 5.5 Assumptions used to construct the beam model for transverse deflections

Assumptions used to construct the beam transverse deflection model
1. The object has one dimension that is large compared to the other two
2. The beam is loaded uniquely in the direction normal to its largest dimension (called a beam)
3. The material properties do not change in the y or z coordinate directions (normal to the long axis: x)
4. The beam is at rest
5. Cross-sections that are planar and normal to the x-axis before loading remain planar and normal to the mid-plane after loading
6. The material is orthotropic and must behave linear elastically at all points in the bar
7. $\sigma_{xx} \gg \sigma_{yy}, \sigma_{zz}$
8. Displacements in the y coordinate direction are small

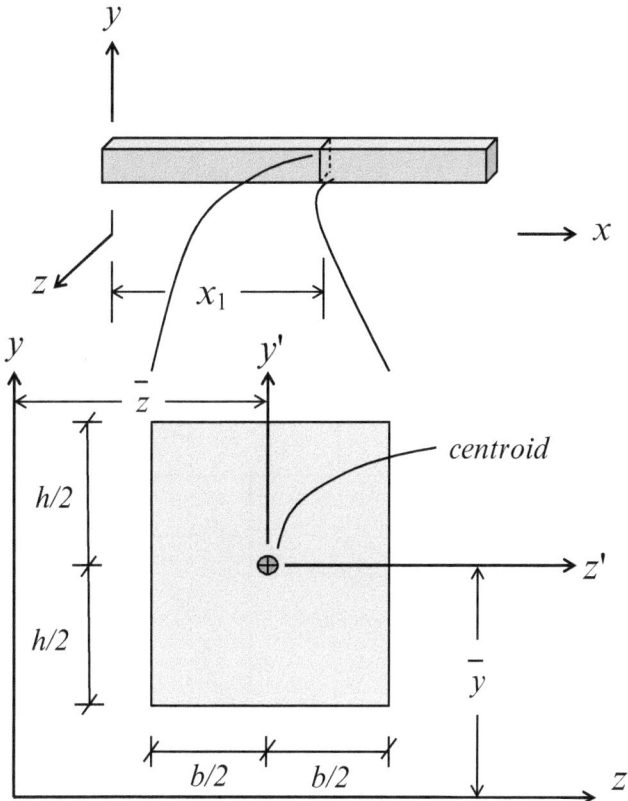

Fig. 5.11 Depiction of a beam with rectangular cross-section

5.2.3   Calculation of the Moment of Inertia

The detailed analysis of a beam of a particular shape will require the calculation of the moment of inertia defined in (5.18). It is clear from that equation that the moment of inertia is in general a function of x, i.e., $I_{zz} = I_{zz}(x)$. The value obtained for a particular coordinate location x_1 in a beam will depend on the shape of the cross-section of the beam at that coordinate location. An example of a beam with a rectangular cross-section is shown in Fig. 5.11, where it is clear from symmetry that the geometric centroid of the rectangle is at the center of the rectangle. Careful integration of (5.18) will reveal that for a rectangle

$$I_{zz} = \int\limits_{-b/2}^{b/2} \int\limits_{-h/2}^{h/2} y^2 \, dy \, dz \Rightarrow I_{zz} = \frac{bh^3}{12} \tag{5.22}$$

Fig. 5.12 Depiction of rectangular cross-section for the purpose of calculating the moment of inertia

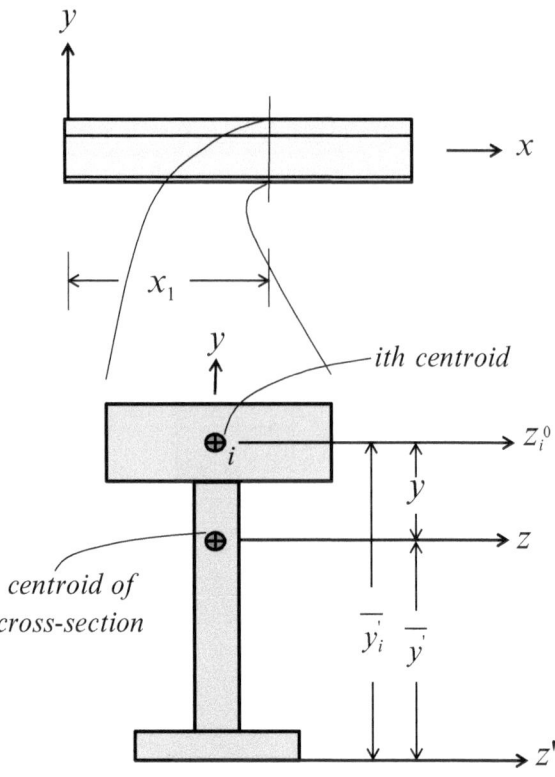

Since tables for cross-section of various shapes can be found readily in a variety of handbooks on the subject, only a simplified shape will be considered herein. Consider a cross-section composed of n rectangles, where an example for $n = 3$ is shown in Fig. 5.12. For purposes of this discussion, it is assumed that we wish to calculate the moment of inertia of the cross-section about the z-axis, which is the centroid of the cross-section. Also shown in the figure are the z-axis for the centroid of the ith rectangle, denoted as z_i^0, and an arbitrarily constructed axis, z'. It is our objective to determine the moment of inertia I_{zz} *about the centroidal axis, z.* However, as will be seen below, it is more convenient to first calculate the moment of inertia about the arbitrary axis, z', due to the fact that it is also necessary to determine the y coordinate of the centroidal axis, $\bar{y}'$. To see how this may be accomplished, first recall the formula for calculating the centroid of the cross-section.

$$\bar{y}' \equiv \frac{1}{A} \sum_{i=1}^{n} \int_{A_i} y' \, dA = \frac{1}{A} \left(\sum_{i=1}^{n} \bar{y}'_i A_i \right) \qquad (5.23)$$

where the subscript i refers to the ith rectangular area within the cross-section.

Next consider the calculation of the moment of inertia about the arbitrary z'-axis:

$$I_{z'z'} = \int_A (y')^2 \, dA = \sum_{i=1}^{n} \int_{A_i} (y')^2 \, dA \qquad (5.24)$$

From Fig. 5.17 it is also apparent that

$$y' = y_i^0 + \bar{y'}_i \qquad (5.25)$$

Substituting (5.25) into (5.24) results in

$$I_{z'z'} = \sum_{i=1}^{n} \int_{A_i} (y_i^0 + \bar{y'}_i)^2 \, dA \Rightarrow$$

$$I_{z'z'} = \sum_{i=1}^{n} \left[\int_{A_i} (y_i^0)^2 \, dA + 2 \int_{A_i} y_i^0 \bar{y'}_i \, dA + \int_{A_i} (\bar{y'})^2 \, dA \right] \qquad (5.26)$$

The first integral in the above equation is $I_{z^0 z^0}$, the moment of inertia of the ith rectangle about its own centroidal axis, given by (5.22). The second integral in (5.26) is zero because the integration is performed with respect to the centroid of the ith rectangle. Therefore, (5.26) may be written as follows:

$$I_{z'z'} = \sum_{i=1}^{n} \left(I_{z_i^0 z_i^0} + \bar{y'}_i^2 A_i \right) \qquad (5.27)$$

where $\bar{y'}_i$ is identical to that employed in (5.23). It can be seen that (5.27) results in the moment of inertia of the cross-section about the arbitrary z'-axis. In order to determine the moment of inertia of the cross-section about the centroidal z-axis, note first that

$$y' = y + \bar{y'} \qquad (5.28)$$

Therefore, substituting the above into (5.24) results in

$$I_{z'z'} = \int_A (y + \bar{y'})^2 \, dA \Rightarrow$$

$$I_{z'z'} = \int_A y^2 \, dA + 2 \int_A y\bar{y'} \, dA + \int_A \bar{y'}^2 \, dA$$

The first integral in the above equation is I_{zz}, and the second integral is zero because $\bar{y} = 0$. Thus, the above may be rearranged and written as follows:

$$I_{zz} = I_{z'z'} - \bar{y'}^2 A \qquad (5.29)$$

The above equation may be used to transform the moment of inertia of the cross-section from the arbitrary z'-axis to the centroidal z-axis, thus completing the calculation of the moment of inertia. The procedure for calculating the moment of inertia about the centroidal axis is reviewed in the following example problem.

Example Problem 5.1
Given: For the cross-section shown below $b = h = 0.10\,\text{m}$

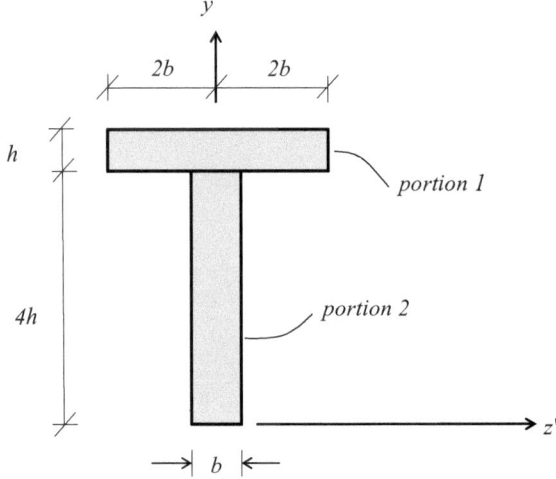

Required

(a) Find the centroid of the cross-section about the z'-axis, $\bar{y}'$
(b) Find the moment of inertia about the z'-axis, $I_{z'z'}$
(c) Find the moment of inertia about the centroid, I_{zz}

Solution

(a) By symmetry $\bar{z} = 0$. To calculate $\bar{y}'$, employ (5.23) as follows:

Portion (i)	$A_i(m^2)$	$\bar{y}'_i(m)$	$\bar{y}'_i A_i(m^3)$
1	0.04	0.45	0.018
2	0.04	0.20	0.008
	$A = 0.08$		$\sum \bar{y}'_i A_i = 0.026$

$$\bar{y}' = \frac{1}{A}\sum \bar{y}'_i A_i = \frac{0.026}{0.08} \Rightarrow \boxed{\bar{y}' = 0.325\,\text{m}} \quad \text{Q.E.D.} \qquad (E5.1.1)$$

(b) Now use (5.27) and (5.28) as follows to find $I_{z'z'}$:

Portion (i)	$A_i(m^2)$	$I_{z^0 z_i^0}(m^4)$	$\overline{y'}_i(m)$	$\overline{y'}_i^2 A_i(m^4)$
1	0.04	0.0000333	0.45	0.0081
2	0.04	0.000533	0.20	0.0016
	$\sum I_{z^0 z_i^0} = 0.0005663$			$\sum \overline{y'}_i^2 A_i = 0.0097$

$$I_{z'z'} = \sum I_{z^0 z_i^0} + \sum \overline{y'}_i^2 A_i = 0.0005663 + 0.0097$$
$$\Rightarrow \boxed{I_{z'z'} = 0.010266\,\mathrm{m}^4} \quad \text{Q.E.D.} \tag{E5.1.2}$$

(c) In order to determine the moment of inertia about the centroid of the cross-section, I_{zz}, use transform formula (5.29) as follows:

$$I_{zz} = I_{z'z'} - \overline{y}'^2 A = 0.010266 - 0.0325^2 \times 0.08$$
$$\Rightarrow \boxed{I_{zz} = 0.001816\,\mathrm{m}^4} \quad \text{Q.E.D.} \tag{E5.1.3}$$

5.2.4   Methods for Obtaining Solutions with the Beam Transverse Deflection Model

Unfortunately, we are not quite finished yet with our model. We still need to develop systematic methods for solving the five equations shown in Table 5.4 for the five unknowns: $V_y = V_y(x)$, $M_z = M_z(x)$, $\sigma_{xx} = \sigma_{xx}(x, y)$, $\varepsilon_{xx} = \varepsilon_{xx}(x, y)$, and $v_0 = v_0(x)$. Before doing that let us make a few observations regarding the equations shown in Table 5.4. These are as follows:

1. Note that the equations are partially coupled, meaning that more than one of the unknowns (in boxes) occurs in several of the equations [except (5.6)]!
2. The equations are all ordinary differential equations, meaning that there is only one independent variable: x (although y is an independent variable, it does not appear in the derivatives).
3. A careful examination of the equations reveals that all of our inputs occur explicitly in the equations in the form of loads: p_y and F_y; geometry: I_{zz} (and L, via the boundary conditions); and material property: E.
4. As a consequence of observation (3), the resulting five predictive equations for $V_y = V_y(x)$, $M_z = M_z(x)$, $\sigma_{xx} = \sigma_{xx}(x, y)$, $\varepsilon_{xx} = \varepsilon_{xx}(x, y)$, and $v_0 = v_0(x)$ that we will obtain by solving the five equations in Table 5.4 will turn out to contain the inputs as well as x and y on the right hand side of the five equations.

Using the above observations, let us now proceed to construct systematic methods for obtaining equations of the form described in observation (4). As it turns out, there are lots of ways of solving the set of five equations. Sometimes one way is easier than another, and this creates lots of confusion for students. We want to minimize this confusion, and for this reason we will develop consistent approaches that will always work.

5.2.4.1 Solution Methods for Statically Determinate Beams

Recall that boundary conditions can be physically applied to the beam in three different ways, as shown in Fig. 5.3. Thus, there can be as few as zero kinematic boundary conditions, and as many as four, depending on how the conditions at the boundary are applied. Because there are only two equations of equilibrium, as described by (5.4), this implies that if there are more than two kinematic boundary conditions, the force and moment reactions at the boundaries cannot be determined simply by using (5.4) a priori. Alternatively, if there are only two kinematic boundary conditions, then (5.4) can be used a priori to determine the force and moment reactions where these kinematic boundary conditions are applied. In the case where one displacement and one rotation boundary condition are applied to the bar, or two displacement boundary conditions are applied to the bar, the problem is called "statically determinate." To restate, this is due to the fact that in this case one can construct a free body diagram of the entire bar and determine the reactions where the displacement and rotation boundary conditions are applied by employing the two equilibrium equations (5.4). In this case, there is a method for solving problems that is distinct from what is appropriate for statically indeterminate problems. The solution method for statically determinate beams is described in Table 5.6.

The above method will result in five explicit equations for the unknowns for $V_y = V_y(x)$, $M_z = M_z(x)$, $\sigma_{xx} = \sigma_{xx}(x, y)$, $\varepsilon_{xx} = \varepsilon_{xx}(x, y)$, and $v_0 = v_0(x)$ as functions of the inputs p_y, F_y, I_{zz}, L, and E and the independent variables, x and y. Once these equations are constructed, they can be used to ensure that the bar is designed in such a way that the design constraints (in this course, fracture and excessive displacements) can be satisfied.

Table 5.6 Systematic method for solving for the transverse displacement in a statically determinate beam

Systematic solution method for transverse displacement in statically determinate beams
Step 1: Solve (5.6) for $V_y = V_y(x)$ using direct integration (one force boundary condition is required)
Step 2: Use $V_y(x)$ obtained in step 1 to solve (5.8) for $M_z = M_z(x)$ (this will require one moment boundary condition)
Step 3: Use $M_z(x)$ obtained in step 2 to obtain $\sigma_{xx} = \sigma_{xx}(x, y)$ using (5.19)
Step 4: Use $\sigma_{xx}(x, y)$ obtained in step 3 to obtain $\varepsilon_{xx} = \varepsilon_{xx}(x, y)$ in (5.12)
Step 5: Use $M_z(x)$ obtained in step 2 to obtain $v_0 = v_0(x)$ using direct integration twice in (5.21) (this will require two boundary conditions of displacement or rotation type)

Example Problem 5.2
Given: The prismatic and homogeneous beam shown below is cantilevered at the left end and has a force F_y applied at the free end $(x = L)$

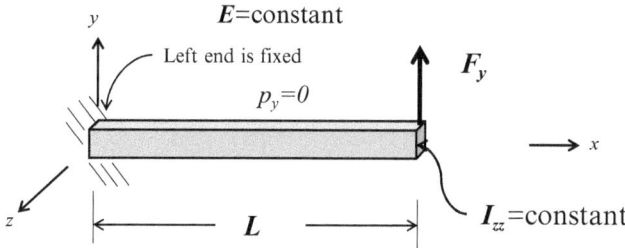

Required

1. Using beam theory, derive an expression for each of the following:

(a) $V_y = V_y(x)$
(b) $M_z = M_z(x)$
(c) $\sigma_{xx} = \sigma_{xx}(x, y, F_y, I_{zz}, L, E)$
(d) $\varepsilon_{xx} = \varepsilon_{xx}(x, y, F_y, I_{zz}, L, E)$
(e) $v_0 = v_0(x, F_y, I_{zz}, L, E)$

2. Plot the results of (a)–(e) on five different graphs: $V_y = V_y(x)$, $M_z = M_z(x)$, $\sigma_{xx} = \sigma_{xx}(x, y)$, $\varepsilon_{xx} = \varepsilon_{xx}(x, y)$, and $v_0 = v_0(x)$
3. Suppose the beam is 10 m long, with $F_y = 10{,}000$ N and is made of A36 steel $(E = 200\,\text{GPa})$, and the maximum allowable displacement of the right end of the bar is 0.2 m, determine the minimum allowable vertical dimension of the bar, h, assuming the cross-section of the beam is square.

Solution

1. (a) Since $p_y = 0$, it follows that, from (5.6)

$$\frac{dV_y}{dx} = -p_y = 0 \Rightarrow \int \frac{dV_y}{dx}\,dx = \int 0\,dx \Rightarrow V_y(x) = C_1 \qquad \text{(E5.2.1)}$$

Now consider the force, F_y, applied at the right end of the bar. Newton's third law may be used to deduce that the boundary condition at the right end is $V_y(x = L) = F_y$. Applying this boundary condition to (E5.2.1) results in

$$\boxed{V_y(x) = F_y} \quad \text{Q.E.D.} \qquad \text{(E5.2.2)}$$

(b) Using (5.8) and (E5.2.2) results in the following:

$$\frac{dM_z}{dx} = -V_y = -F_y \Rightarrow \int \frac{dM_z}{dx}\,dx = -\int F_y\,dx \Rightarrow M_z(x)$$
$$= -F_y x + C_2 \qquad \text{(E5.2.3)}$$

Since the right end of the bar does not have an externally applied moment, it follows that $M_z(x = L) = 0$. Applying this boundary condition to (E5.2.3) results in

$$M_z(x = L) = 0 = -F_yL + C_2 \Rightarrow C_2 = F_yL \tag{E5.2.4}$$

Thus, substituting (E5.2.4) into (E5.2.3) results in the following

$$\boxed{M_z(x) = F_y(L - x)} \quad \text{Q.E.D.} \tag{E5.2.5}$$

(c) Using (5.19) and (E5.2.5), it follows that

$$\sigma_{xx} = -\frac{M_z y}{I_{zz}} \Rightarrow \boxed{\sigma_{xx}(x, y) = \frac{F_y(x - L)y}{I_{zz}}} \quad \text{Q.E.D.} \tag{E5.2.6}$$

(d) Using (5.12) and (E5.2.6) results in the following:

$$\varepsilon_{xx} = -\frac{\sigma_{xx}}{E} \Rightarrow \boxed{\varepsilon_{xx}(x, y) = \frac{F_y(x - L)y}{EI_{zz}}} \quad \text{Q.E.D.} \tag{E5.2.7}$$

(e) Using (5.21) and (E5.2.5) results in the following:

$$\frac{d^2 v_0}{dx^2} = \frac{M_z}{EI_{zz}} = \frac{F_y(L - x)}{EI_{zz}} \Rightarrow \int \frac{d^2 v_0}{dx^2}\, dx = \int \frac{F_y(L - x)}{EI_{zz}}\, dx \Rightarrow$$

$$\frac{dv_0}{dx}(x) = \frac{F_y Lx}{EI_{zz}} - \frac{F_y x^2}{2EI_{zz}} + C_3 \tag{E5.2.8}$$

Applying the boundary condition $dv_0/dx(x = 0) = 0$ to (E5.2.8) implies that $C_3 = 0$. Therefore, (E5.2.8) simplifies and may be solved as follows:

$$\int \frac{dv_0}{dx}\, dx = \int \left[\frac{F_y Lx}{EI_{zz}} - \frac{F_y x^2}{2EI_{zz}} \right] dx \Rightarrow v_0(x) = \frac{F_y Lx^2}{2EI_{zz}} - \frac{F_y x^3}{6EI_{zz}} + C_4 \tag{E5.2.9}$$

Applying the boundary condition $v_0(x = 0) = 0$ implies that $C_4 = 0$, so that (E5.2.9) simplifies to the following:

$$\boxed{v_0(x) = \frac{F_y}{EI_{zz}} \left(\frac{Lx^2}{2} - \frac{x^3}{6} \right)} \quad \text{Q.E.D.} \tag{E5.2.10}$$

2. Plotting (E5.2.2), (E5.2.5), (E5.2.6), (E5.2.7), and (E5.2.10) results in

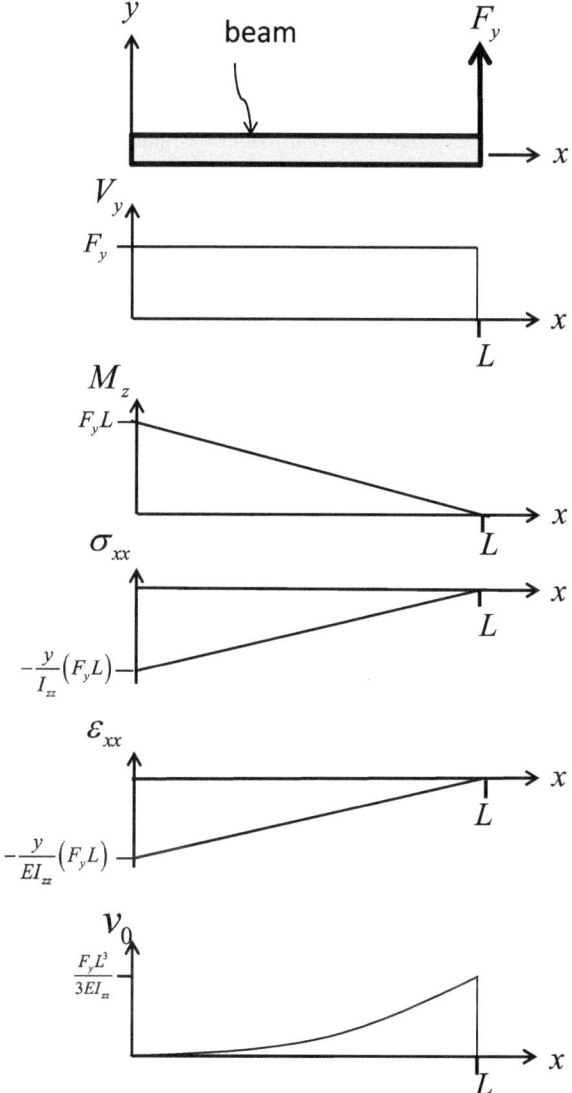

3. The maximum displacement occurs at $x = L$, so that (E5.2.10) gives

$$v_0^{\max} = v_0(x = L) = \frac{F_y}{EI_{zz}}\left(\frac{L^3}{2} - \frac{L^3}{6}\right) \qquad v_0^{\max} = \frac{F_y L^3}{3EI_{zz}} \qquad \text{(E5.2.11)}$$

Applying the values from the problem given result in

$$I_{zz} = h^4/12 = \frac{FL^3}{3Ev_{\max}} = \frac{10,000 \times 10^3}{3 \times 200 \times 10^9 \times 0.2} = 0.833 \times 10^{-4} \Rightarrow$$

$$h = \sqrt[4]{12 \times 0.833 \times 10^{-4}} \Rightarrow \boxed{h = 0.178\,\text{m}} \quad \text{Q.E.D.} \qquad \text{(E5.2.12)}$$

5.2.4.2   Solution Methods for Statically Indeterminate Beams

In the case where at least three displacement and/or rotation boundary conditions are applied to the beam, the problem is statically indeterminate because there are three or more reactions and only two equilibrium equations, thus making it intractable to find the reactions using equilibrium (5.2) by itself. In this case an alternate solution method is preferable. The method is constructed by first substituting (5.21) into (5.8), resulting in the following:

$$\frac{d}{dx}\left(EI_{zz}\frac{d^2v_0}{dx^2}\right) = -V_y \tag{5.30}$$

Equation (5.30) is then substituted into (5.6) to produce

$$\frac{d^2}{dx^2}\left(EI_{zz}\frac{d^2v_0}{dx^2}\right) = p_y \tag{5.31}$$

Equation (5.31) can now be integrated four times, together with four boundary conditions, to obtain $v_0 = v_0(x)$. The remaining unknowns can be found by back substitution into the other equations, as described in Table 5.7.

Table 5.7 Systematic method for solving for the transverse displacement in a statically indeterminate beam

Systematic solution method for transverse displacements in statically indeterminate beams
Step 1: Solve (5.31) for $v_0 = v_0(x)$ using direct integration (this will require four displacement and rotation boundary conditions)
Step 2: Use $v_0(x)$ obtained in step 1 to obtain $M_z = M_z(x)$ using (5.21)
Step 3: Use $M_z(x)$ obtained in step 2 to obtain $V_y = V_y(x)$ using (5.8)
Step 4: Use $M_z(x)$ obtained in step 2 to obtain $\sigma_{xx} = \sigma_{xx}(x,y)$ using (5.19)
Step 5: Use $\sigma_{xx}(x)$ obtained in step 4 to obtain $\varepsilon_{xx} = \varepsilon_{xx}(x)$ using (5.12)

Example Problem 5.3
Given: The prismatic and homogeneous uniaxial bar shown below has a constant applied load per unit length p_y^0

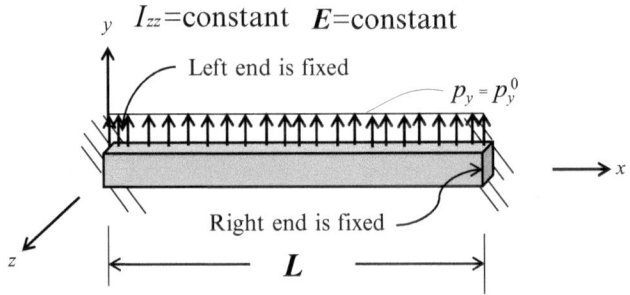

Required

1. Using beam theory, derive an expression for each of the following:

   (a) $v_0 = v_0(x, p_y^0, I_{zz}, L, E)$
   (b) $M_z = M_z(x, p_y^0, I_{zz}, L, E)$
   (c) $V_y = V_y(x, p_y^0, I_{zz}, L, E)$
   (d) $\sigma_{xx} = \sigma_{xx}(x, y, p_y^0, I_{zz}, L, E)$
   (e) $\varepsilon_{xx} = \varepsilon_{xx}(x, y, p_y^0, I_{zz}, L, E)$

2. Plot the results of (a)–(e) on five different graphs: $V_y = V_y(x)$, $M_z = M_z(x)$, $\sigma_{xx} = \sigma_{xx}(x, y)$, $\varepsilon_{xx} = \varepsilon_{xx}(x, y)$, and $v_0 = v_0(x)$
3. Find the reactions at the left and right ends of the beam.

Solution

1. (a) Since $p_y = p_y^0 = $ constant, it follows that, from (5.31)

$$\frac{d^2}{dx^2}\left(EI_{zz}\frac{d^2v_0}{dx^2}\right) = p_y^0 \Rightarrow \int \frac{d^2}{dx^2}\left(EI_{zz}\frac{d^2v_0}{dx^2}\right)dx = \int p_y^0\,dx \Rightarrow$$

$$\frac{d}{dx}\left(EI_{zz}\frac{d^2v_0}{dx^2}\right) = p_y^0 x + C_1 \Rightarrow \int \frac{d}{dx}\left(EI_{zz}\frac{d^2v_0}{dx^2}\right)dx = \int \left(p_y^0 x + C_1\right)dx \Rightarrow$$

$$EI_{zz}\frac{d^2v_0}{dx^2} = p_y^0\frac{x^2}{2} + C_1 x + C_2 \Rightarrow \int EI_{zz}\frac{d^2v_0}{dx^2}\,dx$$

$$= \int \left(p_y^0\frac{x^2}{2} + C_1 x + C_2\right)dx \Rightarrow$$

$$\frac{dv_0}{dx} = \frac{1}{EI_{zz}}\left(p_y^0\frac{x^3}{6} + C_1\frac{x^2}{2} + C_2 x + C_3\right) \Rightarrow$$

$$\tag{E5.3.1}$$

$$\int EI_{zz}\frac{dv_0}{dx}\,dx = \int \left(p_y^0\frac{x^3}{6} + C_1\frac{x^2}{2} + C_2 x + C_3\right)dx \Rightarrow$$

$$v_0(x) = \frac{1}{EI_{zz}}\left(p_y^0\frac{x^4}{24} + C_1\frac{x^3}{6} + C_2\frac{x^2}{2} + C_3 x + C_4\right) \tag{E5.3.2}$$

Now consider the boundary condition $dv_0/dx(x=0) = 0$. It can be seen from (E5.3.2) that

$$C_3 = 0 \tag{E5.3.3}$$

Similarly, consider the boundary condition $v_0(x=0) = 0$. It can be seen from (E5.3.2) that

$$C_4 = 0 \tag{E5.3.4}$$

Next consider the boundary condition $dv_0/dx(x=L) = 0$. It can be seen from (E5.3.1) and (E5.3.3) that

$$0 = \frac{p_y^0 L^3}{6} + \frac{C_1 L^2}{2} + C_2 L \Rightarrow C_2 = -\frac{p_y^0 L^2}{6} - \frac{C_1 L}{2} \tag{E5.3.5}$$

Similarly, consider the boundary condition $v_0(x = L) = 0$. It can be seen from (E5.3.2) and (E5.3.4) that

$$0 = \frac{p_y^0 L^2}{24} + \frac{C_1 L}{6} + \frac{C_2}{2} \Rightarrow C_2 = -\frac{p_y^0 L^2}{12} - \frac{C_1 L}{3} \tag{E5.3.6}$$

Equating (E5.3.5) and (E5.3.6) results in

$$C_1 = -\frac{p_y^0 L}{2} \tag{E5.3.7}$$

Substituting the above into (E5.3.5) results in

$$C_2 = \frac{p_y^0 L^2}{12} \tag{E5.3.8}$$

Thus, substituting (E5.3.3), (E5.3.4), (E5.3.7), and (E5.3.8) into (E5.3.2) gives

$$\boxed{v_0(x) = \frac{p_y^0}{EI_{zz}} \left(\frac{x^4}{24} - \frac{Lx^3}{12} + \frac{L^2 x^2}{24} \right)} \quad \text{Q.E.D.} \tag{E5.3.9}$$

(b) Substituting (E5.3.9) into (5.21) results in the following:

$$M_z(x) = EI_{zz} \frac{d^2 v_0}{dx^2} = p_y^0 \frac{d^2}{dx^2} \left(\frac{x^4}{24} - \frac{Lx^3}{12} + \frac{L^2 x^2}{24} \right) \Rightarrow$$

$$\boxed{M_z(x) = p_y^0 \left(\frac{x^2}{2} - \frac{Lx}{2} + \frac{L^2}{12} \right)} \quad \text{Q.E.D.} \tag{E5.3.10}$$

(c) Substituting (E5.3.10) into (5.8) results in the following:

$$V_y = -\frac{dM_z}{dx} = -\frac{d}{dx} \left[p_y^0 \left(\frac{x^2}{2} - \frac{Lx}{2} + \frac{L^2}{12} \right) \right] \Rightarrow$$

$$\boxed{V_y(x) = p_y^0 \left(\frac{L}{2} - x \right)} \quad \text{Q.E.D.} \tag{E5.3.11}$$

(d) Substituting (E5.3.10) into (5.19) results in the following:

$$\boxed{\sigma_{xx}(x, y) = -\frac{p_y^0 y}{I_{zz}} \left(\frac{x^2}{2} - \frac{Lx}{2} + \frac{L^2}{12} \right)} \quad \text{Q.E.D.} \qquad (E5.3.12)$$

(e) Substituting (E5.3.12) into (5.12) gives

$$\boxed{\varepsilon_{xx}(x, y) = -\frac{p_y^0 y}{EI_{zz}} \left(\frac{x^2}{2} - \frac{Lx}{2} + \frac{L^2}{12} \right)} \quad \text{Q.E.D.} \qquad (E5.3.13)$$

2. Plotting (E5.3.9)–(E5.3.13) results in the following:

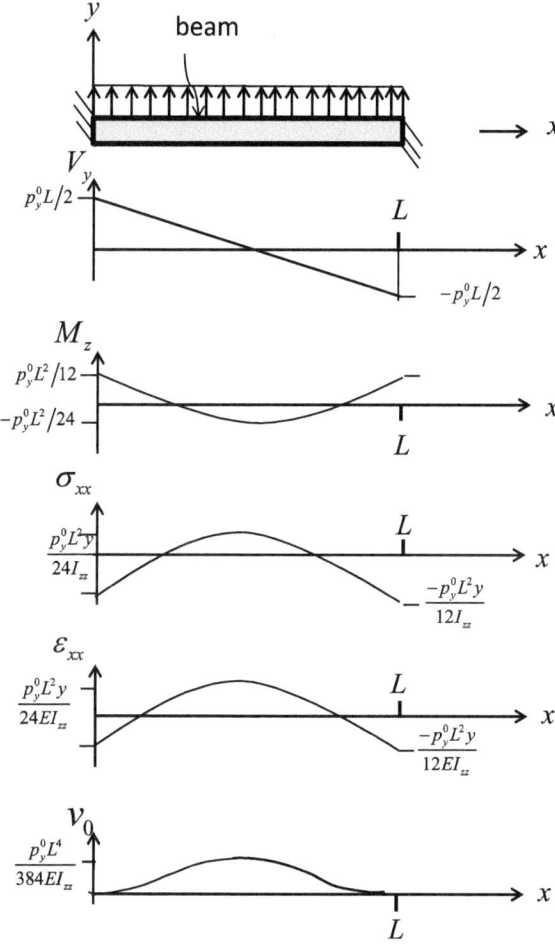

3. It can be seen that the vertical at the left end of the bar, R_L, is equal to the negative of $V_y(x = 0)$. Therefore, using (E5.3.11) results in

$$R_\mathrm{L} = -V_y(x = 0) = -\frac{p_y^0 L}{2} \Rightarrow \boxed{R_\mathrm{L} = -\frac{p_y^0 L}{2}} \quad \text{Q.E.D.} \tag{E5.3.14}$$

Also, the moment reaction at the left end of the bar, M_L, is equal to $M_z(x = L)$. Therefore, using (E5.3.10) results in

$$M_\mathrm{L} = M_z(x = 0) = \frac{p_y^0 L^2}{12} \Rightarrow \boxed{M_\mathrm{L} = \frac{p_y^0 L^2}{12}} \quad \text{Q.E.D.} \tag{E5.3.15}$$

By symmetry, the reactions at the right end are equivalent to the reactions at the left end of the beam.

5.2.4.3   How to Handle Point Forces and Moments

Beams are sometimes subjected to forces and/or moments that are applied over such short distances in the x coordinate direction that for practical purposes they can be considered to be applied at a single point along the x-axis. We term these loads point forces and/or point moments. An example would be the resultant transverse load on a beam caused by the connection of a guide wire along the length of the beam. The inclusion of such forces and/or moments in our model can be accounted for by performing a careful analysis of the kinetics, as described below.

Point Forces

Consider a beam with a point force, $F_y(x_1)$ applied to it, as shown in Fig. 5.13.

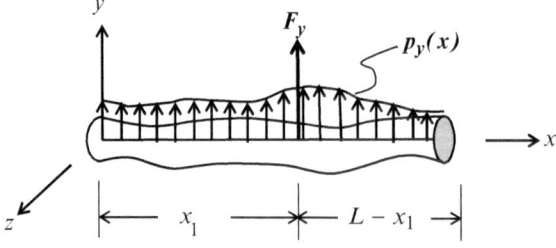

Fig. 5.13 Beam subjected to point force, F_y

Now, suppose that the bar is cut normal to the x-axis at coordinate location x_1^+, just to the right of the location of the force, F_y, as shown in Fig. 5.14.

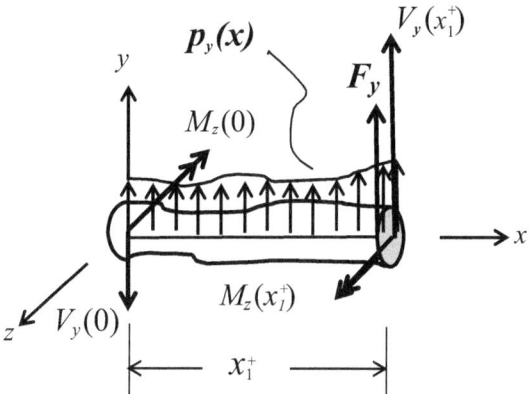

Fig. 5.14 Free body diagram of a beam cut to the right of load, F_y

In this case, summing forces in the y direction will result in the following:

$$\sum F_y = 0 \Rightarrow V_y(x_1^+) - V_y(0) + F_y + \int_0^{x_1^+} p_y(x)\,dx = 0 \Rightarrow$$

$$V_y(x_1^+) = V_y(0) - F_y - \int_0^{x_1^+} p_y\,dx \tag{5.32}$$

On the other hand, if the second cut is made just to the left of where the force F_y is applied, the free body diagram is identical to that shown in Fig. 5.10, with the necessary result that

$$V_y(x_1^-) = V_y(0) - \int_0^{x_1^-} p_y\,dx \tag{5.33}$$

It therefore follows that since the coordinate locations x_1^- and x_1^+ can be made arbitrarily close, there must be a jump discontinuity in $V_y(x)$ at the coordinate location, x_1, where the force, F_y, is applied. It should also be apparent by comparing the results obtained in (5.32) and (5.33) that if the force, F_y, is applied in the positive x direction, the jump is negative and equal in magnitude to F_y. If the force, F_y, is applied in the negative x direction, then the jump is positive and equal in magnitude to F_y. These results are summarized in Table 5.8.

Table 5.8 How to handle point forces applied to beams

How to handle a point force, F_y, applied at coordinate location $x = x_1$

1. If F_y is in the positive x direction, decrease $V_y(x_1)$ by F_y
2. If F_y is in the negative x direction, increase $V_y(x_1)$ by F_y

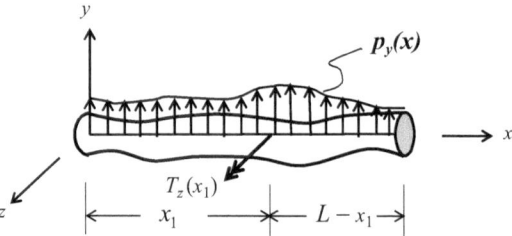

Fig. 5.15 Beam subjected to point moment, T_z

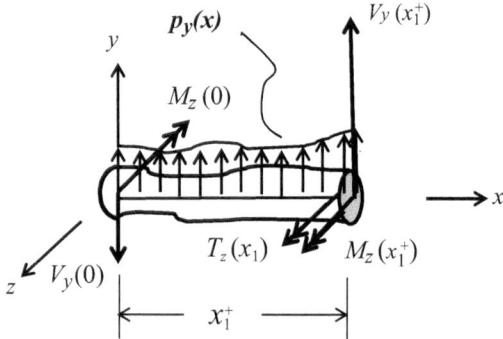

Fig. 5.16 Free body diagram of a beam cut to the right of moment, $T_z(x_1)$

Point Moments

Consider a beam with a point moment applied to it, as shown in Fig. 5.15.

Now, suppose that the bar is cut normal to the x-axis at coordinate location x_1^+, just to the right of the location of the moment, $T_z(x_1)$, as shown in Fig. 5.16.

In this case, summing moments about the z-axis will result in the following:

$$\sum M_z = 0 \Rightarrow M_z(x_1^+) - M_z(0) + T_z(x_1) + \int_0^{x_1^+} \alpha x_1 p_y(x)\, dx = 0 \Rightarrow$$

$$M_z(x_1^+) = M_z(0) - T_z(x_1) - \alpha x_1 \int_0^{x_1^+} p_y\, dx \qquad (5.34)$$

where, as before α is an arbitrary constant such that $0 \leq \alpha \leq 1$. On the other hand, if the second cut is made just to the left of where the moment, T_z, is applied, the free body diagram is similar to that shown in Fig. 5.7, with the necessary result that

$$M_z(x_1^-) = M_z(0) - \alpha x_1 \int_0^{x_1^-} p_y \, dx \tag{5.35}$$

It therefore follows that since the coordinate locations x_1^- and x_1^+ can be made arbitrarily close, there must be a jump discontinuity in $M_z(x)$ at the coordinate location, x_1, where the moment, $T_z(x_1)$, is applied. It should also be apparent by comparing the results obtained in (5.34) and (5.35) that if the moment, $T_z(x_1)$, is applied in the positive z direction, the jump is negative and equal in magnitude to $M_z(x_1)$. If the moment, $T_z(x_1)$, is applied in the negative z direction, then the jump is positive and equal in magnitude to $T_z(x_1)$. These results are summarized in Table 5.9.

Table 5.9 How to handle point moments applied to beams

How to handle a point moment, T_z, applied at coordinate location $x = x_1$
1. If T_z is in the positive z direction, decrease $M_z(x_1)$ by T_z
2. If T_z is in the negative z direction, increase $M_z(x_1)$ by T_z

Example Problem 5.4
Given: The prismatic and homogeneous beam has point forces and point moments as shown below

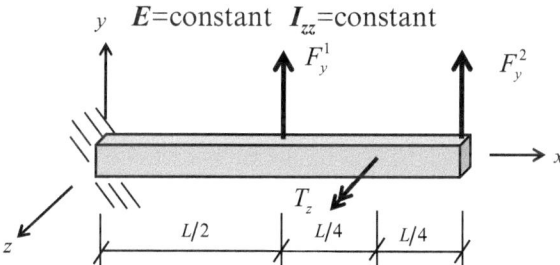

Required

1. Using beam theory, derive an expression for each of the following.

(a) $V_y = V_y(x, F_y^1, F_y^2, L)$
(b) $M_z = M_z(x, F_y^1, F_y^2, T_z, L)$
(c) $\sigma_{xx} = \sigma_{xx}(x, y, F_y^1, F_y^2, T_z, L, E, I_{zz})$
(d) $\varepsilon_{xx} = \varepsilon_{xx}(x, y, F_y^1, F_y^2, T_z, L, E, I_{zz})$
(e) $v_0 = v_0(x, F_y^1, F_y^2, T_z, L, E, I_{zz})$

2. Plot the results of (a)–(e) on five different graphs: $V_y = V_y(x)$, $M_z = M_z(x)$, $\sigma_{xx} = \sigma_{xx}(x)$, $\varepsilon_{xx} = \varepsilon_{xx}(x)$, $v_0 = v_0(x)$

Solution

1. Since the problem is statically determinate, use Table 5.6.

   (a) Solve (5.6) for $V_y = V_y(x)$ as follows:

$$\frac{dV_y}{dx} = -p_y^0 = 0 \Rightarrow V_y = C_1 \quad 0 \leq x \leq L/2$$
$$V_y = C_2 \quad L/2 \leq x \leq L \tag{E5.4.1}$$

Now apply the boundary condition $V_y(x = L) = F_y^2$. Thus, using Table 5.8, it follows that

$$\boxed{\begin{aligned} V_y &= F_y^1 + F_y^2 & 0 \leq x \leq L/2 \\ V_y &= F_y^2 & L/2 \leq x \leq L \end{aligned}} \quad \text{Q.E.D.} \tag{E5.4.2}$$

   (b) Solve (5.8) for $M_z = M_z(x)$ as follows:

$$\frac{dM_z}{dx} = -V_y \Rightarrow \int \frac{dM_z}{dx} dx = -\int (F_y^1 + F_y^2) dx \quad 0 \leq x \leq L/2 \Rightarrow$$
$$M_z(x) = -(F_y^1 + F_y^2)x + C_3 \quad 0 \leq x \leq L/2 \tag{E5.4.3}$$

Also

$$\frac{dM_z}{dx} = -V_y \Rightarrow \int \frac{dM_z}{dx} dx = -\int F_y^2 dx \quad L/2 \leq x \leq L \Rightarrow$$
$$M_z(x) = -F_y^2 x + C_4 \quad L/2 \leq x \leq 3L/4 \tag{E5.4.4a}$$

$$M_z(x) = -F_y^2 x + C_5 \quad 3L/4 \leq x \leq L \tag{E5.4.4b}$$

Next, apply the boundary condition $M_z(x = L) = 0$ and Table 5.9 to obtain

$$M_z(x = L) = 0 = -F_y^2 L + C_5 \Rightarrow C_5 = F_y^2 L \tag{E5.4.5}$$

Thus, substituting (E5.4.5) into (E5.4.4b) and employing Table 5.9 with (E5.4.4a) will result in

$$\boxed{M_z(x) = F_y^2(L - x) + T_z \quad L/2 \leq x \leq 3L/4} \tag{E5.4.6a}$$

$$\boxed{M_z(x) = F_y^2(L - x) \quad 3L/4 \leq x \leq L} \quad \text{Q.E.D.} \tag{E5.4.6b}$$

A matching condition must be applied to account for the fact that the moment must be continuous at $x = L/2$. Applying this condition to (E5.4.3) and (E5.4.6a) will result in

$$-(F_y^1 + F_y^2) \times L/2 + C_3 = F_y^2(L - L/2) + T_z \Rightarrow C_3 = \frac{F_y^1 L}{2} + F_y^2 L + T_z$$

$$(E5.4.7)$$

Finally, substituting the above result into (E5.4.3) gives

$$\boxed{M_z(x) = F_y^1 \left(\frac{L}{2} - x\right) + F_y^2(L - x) + T_z \quad 0 \le x \le L/2}\quad \text{Q.E.D.} \qquad (E5.4.8)$$

(c) Solve (5.19) for $\sigma_{xx} = \sigma_{xx}(x)$ using the results obtained above as follows:

$$\boxed{\sigma_{xx} = -\frac{M_z y}{I_{zz}} \Rightarrow \sigma_{xx}(x, y) = -\frac{y}{I_{zz}}\left[F_y^1 \left(\frac{L}{2} - x\right) + F_y^2(L - x) + T_z\right] 0 \le x \le L/2}$$

$$\boxed{\sigma_{xx}(x, y) = -\frac{y}{I_{zz}}[F_y^2(L - x) + T_z]\ L/2 \le x \le 3L/4}$$

$$\boxed{\sigma_{xx}(x, y) = -\frac{y}{I_{zz}}F_y^2(L - x)\ 3L/4 \le x \le L}\quad \text{Q.E.D.}$$

$$(E5.4.9)$$

(d) Solve (5.12) for $\varepsilon_{xx} = \varepsilon_{xx}(x)$ as follows:

$$\boxed{\varepsilon_{xx}(x, y) = -\frac{y}{EI_{zz}}\left[F_y^1 \left(\frac{L}{2} - x\right) + F_y^2(L - x) + T_z\right] 0 \le x \le L/2}$$

$$\boxed{\varepsilon_{xx}(x, y) = -\frac{y}{EI_{zz}}[F_y^2(L - x) + T_z]\ L/2 \le x \le 3L/4}$$

$$\boxed{\varepsilon_{xx}(x, y) = -\frac{y}{EI_{zz}}F_y^2(L - x)\ 3L/4 \le x \le L}\quad \text{Q.E.D.} \qquad (E5.4.10)$$

(e) Solve for $v_0 = v_0(x)$ by substituting (E5.4.6) and (E5.4.8) into (5.21) as follows:

$$\frac{d^2 v_0}{dx^2} = \frac{M_z}{EI_{zz}} \Rightarrow$$

$$\int \frac{d^2 v_0}{dx^2}\,dx = \frac{1}{EI_{zz}}\int \left[F_y^1 \left(\frac{L}{2} - x\right) + F_y^2(L - x) + T_z\right]dx \Rightarrow$$

$$\frac{dv_0}{dx} = \frac{1}{EI_{zz}}\left[F_y^1 \left(\frac{Lx}{2} - \frac{x^2}{2}\right) + F_y^2 \left(Lx - \frac{x^2}{2}\right) + T_z x\right] + C_6 \qquad (E5.4.11)$$

Applying the boundary condition $dv_0/dx(x = 0) = 0$ implies that $C_6 = 0$. Integrating (E5.4.11) a second time therefore gives

$$v_0(x) = \frac{1}{EI_{zz}}\left[F_y^1 \left(\frac{Lx^2}{4} - \frac{x^3}{6}\right) + F_y^2 \left(\frac{Lx^2}{2} - \frac{x^3}{6}\right) + T_z \frac{x^2}{2}\right] + C_7$$

Applying the boundary condition $v(x = 0) = 0$ implies that $C_7 = 0$. The above therefore simplifies to the following:

$$v_0(x) = \frac{1}{EI_{zz}}\left[F_y^1\left(\frac{Lx^2}{4} - \frac{x^3}{6}\right) + F_y^2\left(\frac{Lx^2}{2} - \frac{x^3}{6}\right) + T_z\frac{x^2}{2}\right] \quad 0 \le x \le L/2 \quad \text{Q.E.D.}$$

$$(E5.4.12)$$

Similarly, integrating (E5.4.6a) results in

$$\frac{dv_0}{dx}(x) = \frac{1}{EI_{zz}}\left[F_y^2\left(Lx - \frac{x^2}{2}\right) + T_z x\right] + C_8 \qquad (E5.4.13)$$

Since the slope of the beam must be continuous, (E5.4.11) and (E5.4.12) may be equated at $x = L/2$, implying that

$$C_8 = \frac{F_y^1 L^2}{8}$$

Equation (E5.4.13) may therefore be written as follows:

$$\frac{dv_0}{dx}(x) = \frac{1}{EI_{zz}}\left[F_y^2\left(Lx - \frac{x^2}{2}\right) + T_z x\right] + \frac{F_y^1 L^2}{8} \qquad L/2 \le x \le 3L/4 \qquad (E5.4.14)$$

Integrating the above equation once again results in

$$v_0(x) = \frac{1}{EI_{zz}}\left[F_y^2\left(\frac{Lx^2}{2} - \frac{x^3}{6}\right) + T_z\frac{x^2}{2}\right] + \frac{F_y^1 L^2}{8}x + C_9 \qquad (E5.4.15)$$

Since the displacement of the beam must be continuous, (E5.4.12) and (E5.4.15) may be equated at $x = L/2$, implying that

$$C_9 = \frac{5F_y^1 L^3}{48}$$

Equation (E5.4.13) may therefore be written as follows:

$$v_0(x) = \frac{1}{EI_{zz}}\left[F_y^2\left(\frac{Lx^2}{2} - \frac{x^3}{6}\right) + T_z\frac{x^2}{2}\right] + \frac{5F_y^1 L^2}{48}x \quad L/2 \le x \le 3L/4 \quad \text{Q.E.D.} \quad (E5.4.16)$$

Integrating (E5.4.6b) results in

$$\frac{dv_0}{dx}(x) = \frac{1}{EI_{zz}}\left[F_y^2\left(Lx - \frac{x^2}{2}\right) + C_{10}\right] \tag{E5.4.17}$$

Since the slope of the beam must be continuous everywhere, (E5.4.14) and (E5.4.17) may be equated at $x = 3L/4$, implying that

$$C_{10} = \frac{F_y^1 L^2}{8} + \frac{3T_z L}{4}$$

Substituting the above into (E5.4.17) therefore gives

$$\frac{dv_0}{dx}(x) = \frac{1}{EI_{zz}}\left[F_y^2\left(Lx - \frac{x^2}{2}\right) + \frac{5F_y^1 L^2}{48} + \frac{3T_z L}{4}\right]$$

Integrating the above once more gives

$$v_0(x) = \frac{1}{EI_{zz}}\left[F_y^2\left(\frac{Lx^2}{2} - \frac{x^3}{6}\right) + \frac{5F_y^1 L^2 x}{48} + \frac{3T_z Lx}{4} + C_{11}\right]$$

Since the displacement of the beam must be everywhere continuous, the above may be equated to (E5.4.16) at $x = 3L/4$, implying that

$$C_{11} = -\frac{9T_z L^2}{32}$$

Thus, finally

$$\boxed{v_0(x) = \frac{1}{EI_{zz}}\left[F_y^2\left(\frac{Lx^2}{2} - \frac{x^3}{6}\right) + \frac{5F_y^1 L^2 x}{48} + \frac{3T_z Lx}{4} - \frac{9T_z L^2}{32}\right] \quad 3L/4 \le x \le L} \quad \text{Q.E.D.}$$

$$\tag{E5.4.18}$$

2. Plotting (E5.4.2), (E5.4.6), (E5.4.8)–(E5.4.10), (E5.4.12), (E5.4.16), and (E5.4.18) results in the following graphs.

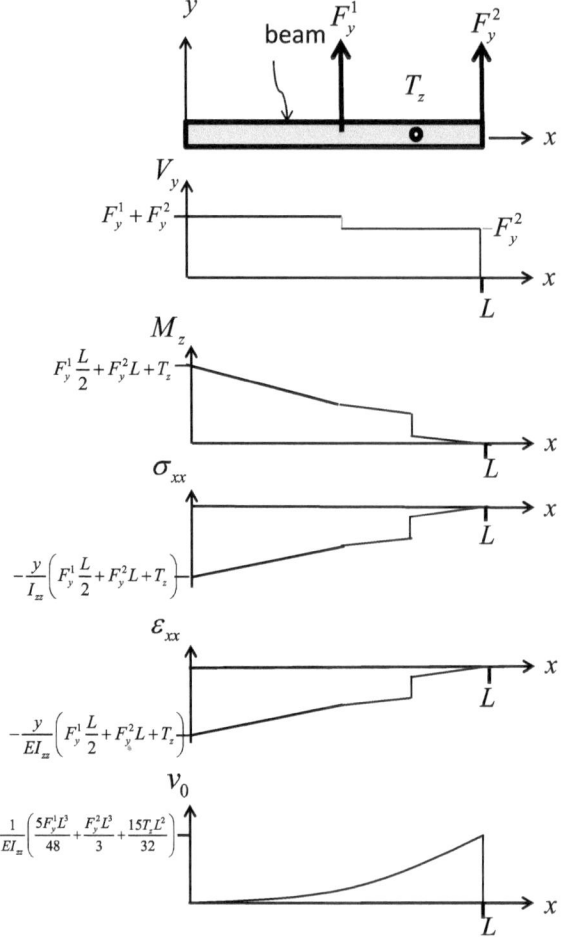

For the reader who is interested in obtaining solutions to more complicated beams problems, see Roark et al. (1975).

5.2.5   A Model for Beam Shear

As described in the previous section, beams are in general subjected to shearing forces, $V_y = V_y(x)$, normal to their long axis, as defined by (5.1), and this shear force can be determined using differential equation of equilibrium (5.6). It is thus clear that there is shear stress, $\sigma_{xy} = \sigma_{xy}(x, y)$, that produces the resultant shear force, V_y, in accordance with (5.1). In order to visualize this shearing stress, first recall from (2.6), obtained by summing moments at a point and employing Newton's first law, that the following is true at all points in the beam.

$$\sigma_{yx} = \sigma_{xy} \tag{5.36}$$

where σ_{yx} is the shear stress in the x direction on the horizontal plane as shown in Fig. 5.17.

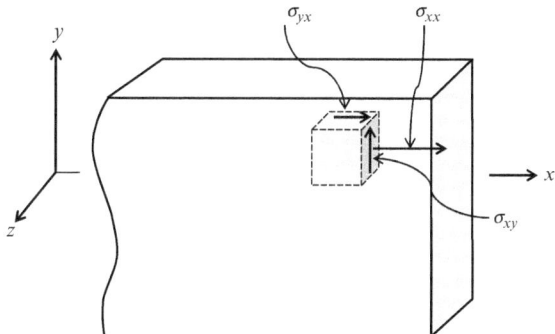

Fig. 5.17 Shear stresses on horizontal and vertical planes in a beam

The fact that there are shear stresses on the horizontal plane in the beam due to externally applied loads can be seen by performing a simple thought experiment. Suppose that two yardsticks are stacked on top of each other, and a beam is subjected to a vertical loading, as shown in Fig. 5.18. Unless the yardsticks are nailed or glued together, the two yardsticks will be seen to slide with respect to one another along the horizontal plane, as shown in the figure. Thus, if the yardsticks are made contiguous, shear stresses will accrue on the horizontal (and associated vertical) planes.

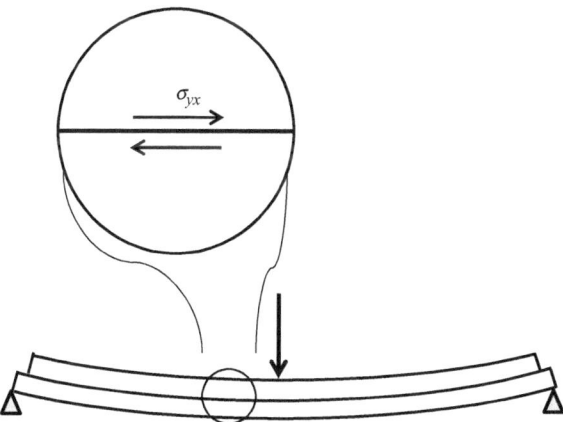

Fig. 5.18 Sliding along the horizontal plane when two yardsticks are stacked and loaded vertically

From the above thought experiment, it is clear that the shear stress resulting from transverse loading in a beam can have a significant deleterious effect on the performance of the beam. Thus, it is important to develop a model for predicting this component of stress in beams.

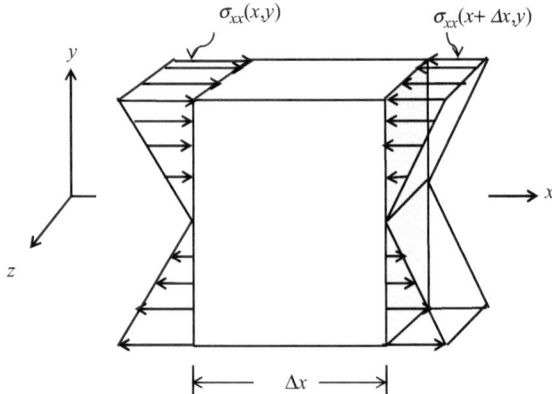

Fig. 5.19 Free body diagram of a section of a beam, showing normal stress components σ_{xx}

In order to construct such a model, consider the free body diagram of a section of a beam, as shown in Fig. 5.19.

Now suppose that the free body diagram shown above is cut on a horizontal plane, as shown in Fig. 5.20 below.

Summing forces in the x direction will result in the following:

$$\sum F_x = 0 = \int \sigma_{xx}(x + \Delta x, y)\, \mathrm{d}A - \int \sigma_{xx}(x, y)\, \mathrm{d}A + \sigma_{yx}(x, y)b\Delta x \qquad (5.37)$$

The above may be rearranged using (5.19) to produce the following form

$$\sigma_{yx}b = \frac{1}{\Delta x}\left[\int \frac{M_z(x + \Delta x)y}{I_{zz}}\,\mathrm{d}A - \int \frac{M_z(x)y}{I_{zz}}\,\mathrm{d}A\right] = \frac{1}{I_{zz}}\left[\frac{M_z(x + \Delta x) - M_z(x)}{\Delta x}\right]\int y\,\mathrm{d}A$$

Taking the limit as Δx approaches zero thus results in the following formula

$$\sigma_{yx}(x, y) = \frac{\mathrm{d}M_z}{\mathrm{d}x} \times \frac{Q(y)}{I_{zz}b(y)} \qquad (5.38)$$

where

$$Q(y) \equiv \int y\,\mathrm{d}A \qquad (5.39)$$

is the integral taken over the gray colored area shown in Fig. 5.20. Thus, substituting (5.8) and (5.36) into (5.38) results in

$$\sigma_{xy}(x, y) = -\frac{V(x)Q(y)}{I_{zz}b(y)} \qquad (5.40)$$

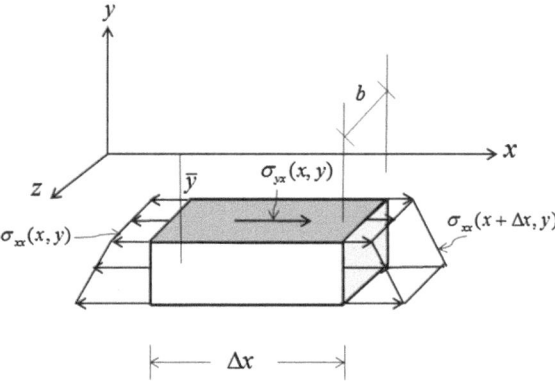

Fig. 5.20 Free body diagram depicting shear stress σ_{yx}

The above is a rather complicated formula to use due to the definition of $Q(y)$ given in (5.39). However, careful manipulation of (5.39) will show that $Q(y)$ may be evaluated using the following equivalent formula

$$Q(y) = A_G \bar{y}_G \qquad (5.41)$$

where A_G is the area of the shaded region in Fig. 5.20, and $\bar{y}_G$ is the distance from the centroid of the shaded area to the z-axis.

Equation (5.40) may be used to predict the shear stress on vertical planes in beams. It should be noted, however, that the value of $Q(y)$, due to definition (5.39) is always zero at the top and bottom of a beam, so that the normal stress, σ_{xx}, is a maximum where the shear stress, σ_{xy}, is zero. Furthermore, the shear stress, σ_{xy}, attains its maximum at the coordinate location $y = 0$, which is precisely where the normal stress, σ_{xx}, is zero. Finally, practice indicates that the maximum normal stress, σ_{xx}, in beams is almost always an order of magnitude larger than the maximum shear stress, σ_{xy}, so that this component of stress is generally not significant for purposes of predicting failure in beams.

5.3   Assignments

PROBLEM 5.1
GIVEN: The definition of a beam.
REQUIRED: Locate a beam either on the university campus or in the local area, describe it (meaning loads, geometry, and material properties), and include a photo of it.

PROBLEM 5.2
GIVEN: The beam shown below is homogeneous, prismatic and has an evenly distributed axial load of $p_y = p_y^0 =$ constant applied along its length as shown.

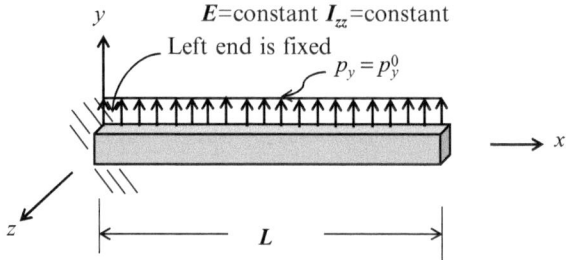

REQUIRED

1. Using beam theory, derive an expression for each of the following:

(a) $V_y = V_y(x, p_y^0, L)$
(b) $M_z = M_z(x, p_y^0, L)$
(c) $\sigma_{xx} = \sigma_{xx}(x, y, p_y^0, I_{zz}, L, E)$
(d) $\varepsilon_{xx} = \varepsilon_{xx}(x, y, p_y^0, I_{zz}, L, E)$
(e) $v_0 = v_0(x, p_y^0, I_{zz}, L, E)$

2. Plot the results of (a)–(e) on five different graphs: $V_y = V_y(x)$, $M_z = M_z(x)$, $\sigma_{xx} = \sigma_{xx}(x)$, $\varepsilon_{xx} = \varepsilon_{xx}(x)$, $v_0 = v_0(x)$ (for a given value of the input loads, geometry, and material properties).

PROBLEM 5.3

GIVEN: A beam has the cross-section shown below.

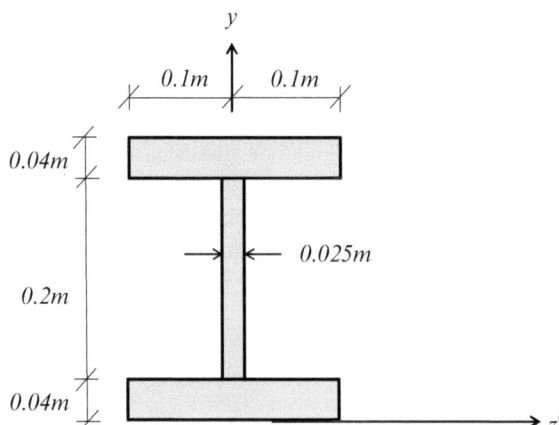

REQUIRED: 1. Determine $\bar{y}'$, $I_{z'z'}$, and I_{zz}, and for the cross-section.

PROBLEM 5.4

GIVEN: The simply supported beam shown below is homogeneous, prismatic and has a distributed axial load of $p_y = p_y^0 x$ applied along its length as shown.

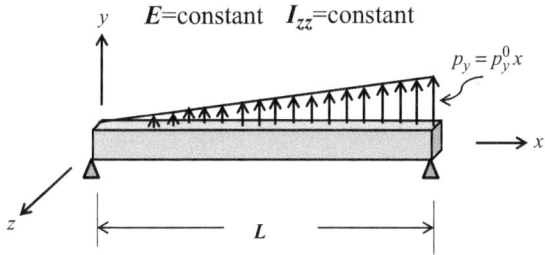

REQUIRED

1. Using beam theory, derive an expression for each of the following:

   (a) $V_y = V_y(x, p_y^0, L)$
   (b) $M_z = M_z(x, p_y^0, L)$
   (c) $\sigma_{xx} = \sigma_{xx}(x, y, p_y^0, I_{zz}, L, E)$
   (d) $\varepsilon_{xx} = \varepsilon_{xx}(x, y, p_y^0, I_{zz}, L, E)$
   (e) $v_0 = v_0(x, p_y^0, I_{zz}, L, E)$

2. Plot the results of (a)–(e) on five different graphs: $V_y = V_y(x)$, $M_z = M_z(x)$, $\sigma_{xx} = \sigma_{xx}(x)$, $\varepsilon_{xx} = \varepsilon_{xx}(x)$, $v_0 = v_0(x)$ (for a given value of the input loads, geometry, and material properties).

3. Find the vertical reactions at the left and right ends of the beam.

PROBLEM 5.5

GIVEN: The double cantilevered beam shown below is homogeneous, prismatic and has a distributed axial load of $p_y = p_x^0 x$ applied along its length as shown.

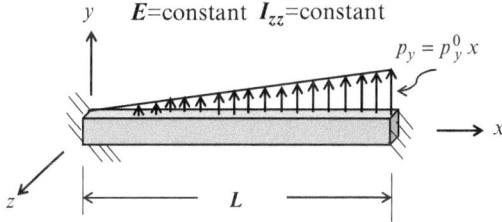

REQUIRED

1. Using beam theory, derive an expression for each of the following:

   (a) $V_y = V_y(x, p_y^0, L)$
   (b) $M_z = M_z(x, p_y^0, L)$
   (c) $\sigma_{xx} = \sigma_{xx}(x, y, p_y^0, I_{zz}, L, E)$
   (d) $\varepsilon_{xx} = \varepsilon_{xx}(x, y, p_y^0, I_{zz}, L, E)$
   (e) $v_0 = v_0(x, p_y^0, I_{zz}, L, E)$

2. Plot the results of (a)–(e) on five different graphs: $V_y = V_y(x)$, $M_z = M_z(x)$, $\sigma_{xx} = \sigma_{xx}(x)$, $\varepsilon_{xx} = \varepsilon_{xx}(x)$, $v_0 = v_0(x)$ (for a given value of the input loads, geometry, and material properties).
3. Find the vertical reactions at the left and right ends of the beam.

PROBLEM 5.6

GIVEN: The simply supported beam shown below is homogeneous, prismatic and has an evenly distributed load per unit length p_y^0 and a point force $F_y = p_y^0 L$ applied at $x = L/2$ as shown.

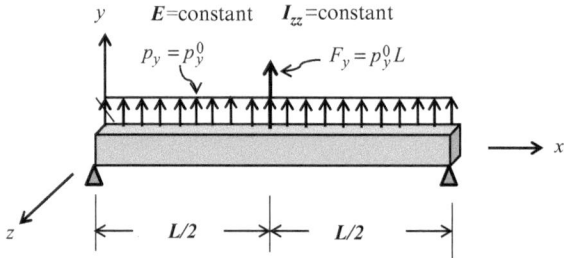

REQUIRED

1. Using beam theory, derive an expression for each of the following:

   (a) $V_y = V_y(x, p_y^0, L)$
   (b) $M_z = M_z(x, p_y^0, L)$
   (c) $\sigma_{xx} = \sigma_{xx}(x, y, p_y^0, I_{zz}, L, E)$
   (d) $\varepsilon_{xx} = \varepsilon_{xx}(x, y, p_y^0, I_{zz}, L, E)$
   (e) $v_0 = v_0(x, p_y^0, I_{zz}, L, E)$

2. Plot the results of (a)–(e) on five different graphs: $V_y = V_y(x)$, $M_z = M_z(x)$, $\sigma_{xx} = \sigma_{xx}(x)$, $\varepsilon_{xx} = \varepsilon_{xx}(x)$, $v_0 = v_0(x)$ (for a given value of the input loads, geometry, and material properties).
3. Find the vertical reactions at the left and right ends of the beam.

PROBLEM 5.7

GIVEN: The cantilevered beam shown below is homogeneous, prismatic and has a point force F_y applied at $x = L/2$ as shown.

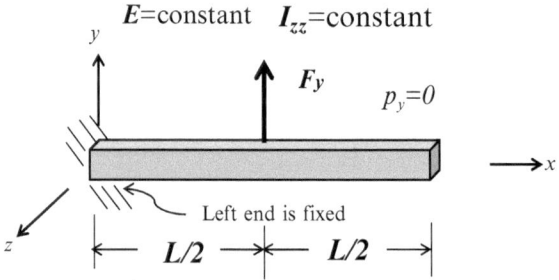

REQUIRED

1. Using beam theory, derive an expression for each of the following:

   (a) $V_y = V_y(x, F_y, I_{zz}, L, E)$
   (b) $M_z = M_z(x, F_y, I_{zz}, L, E)$
   (c) $\sigma_{xx} = \sigma_{xx}(x, y, F_y, I_{zz}, L, E)$
   (d) $\varepsilon_{xx} = \varepsilon_{xx}(x, y, F_y, I_{zz}, L, E)$
   (e) $v_0 = v_0(x, F_y, I_{zz}, L, E)$

2. Plot the results of (a)–(e) on five different graphs: $V_y = V_y(x)$, $M_z = M_z(x)$, $\sigma_{xx} = \sigma_{xx}(x)$, $\varepsilon_{xx} = \varepsilon_{xx}(x)$, $v_0 = v_0(x)$ (for a given value of the input loads, geometry, and material properties).
3. Find the maximum normal stress in the beam and give its location.

PROBLEM 5.8
GIVEN: The simply supported beam shown below is homogeneous, prismatic and has a point force F_y applied at $x = L/2$ as shown.

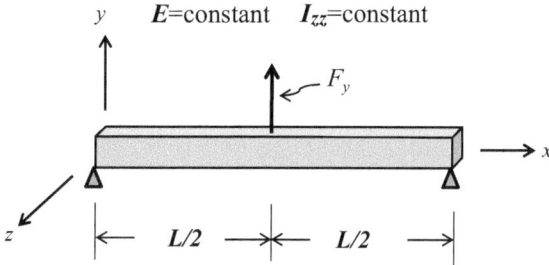

REQUIRED

1. Using beam theory, derive an expression for each of the following:

   (a) $V_y = V_y(x, F_y, I_{zz}, L, E)$
   (b) $M_z = M_z(x, F_y, I_{zz}, L, E)$
   (c) $\sigma_{xx} = \sigma_{xx}(x, y, F_y, I_{zz}, L, E)$
   (d) $\varepsilon_{xx} = \varepsilon_{xx}(x, y, F_y, I_{zz}, L, E)$
   (e) $v_0 = v_0(x, F_y, I_{zz}, L, E)$

2. Plot the results of (a)–(e) on five different graphs: $V_y = V_y(x)$, $M_z = M_z(x)$, $\sigma_{xx} = \sigma_{xx}(x)$, $\varepsilon_{xx} = \varepsilon_{xx}(x)$, $v_0 = v_0(x)$ (for a given value of the input loads, geometry, and material properties)
3. Find the vertical reactions at the left and right ends of the beam.

PROBLEM 5.9
GIVEN: The simply supported beam shown below is homogeneous, prismatic and has a point moment T_z applied at $x = L/2$ as shown.

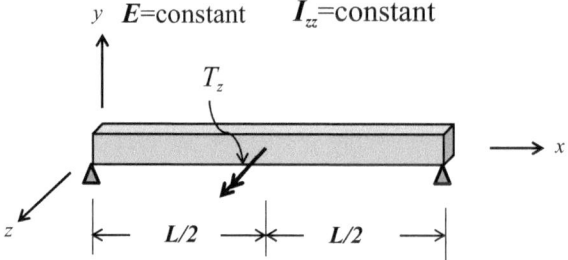

y $E=\text{constant}$ $I_{zz}=\text{constant}$

T_z

x

z

$L/2$ $L/2$

REQUIRED

1. Using beam theory, derive an expression for each of the following:

   (a) $V_y = V_y(x, T_z, I_{zz}, L, E)$
   (b) $M_z = M_z(x, T_z, I_{zz}, L, E)$
   (c) $\sigma_{xx} = \sigma_{xx}(x, y, T_z, I_{zz}, L, E)$
   (d) $\varepsilon_{xx} = \varepsilon_{xx}(x, y, T_z, I_{zz}, L, E)$
   (e) $v_0 = v_0(x, T_z, I_{zz}, L, E)$

2. Plot the results of (a)–(e) on five different graphs: $V_y = V_y(x)$, $M_z = M_z(x)$, $\sigma_{xx} = \sigma_{xx}(x)$, $\varepsilon_{xx} = \varepsilon_{xx}(x)$, $v_0 = v_0(x)$ (for a given value of the input loads, geometry, and material properties).
3. Find the vertical reactions at the left and right ends of the beam.

PROBLEM 5.10
Given: Cantilever beam shown below is made of two pieces of Douglas Fir that are joined together by nails that are capable of carrying 100 N in shear.

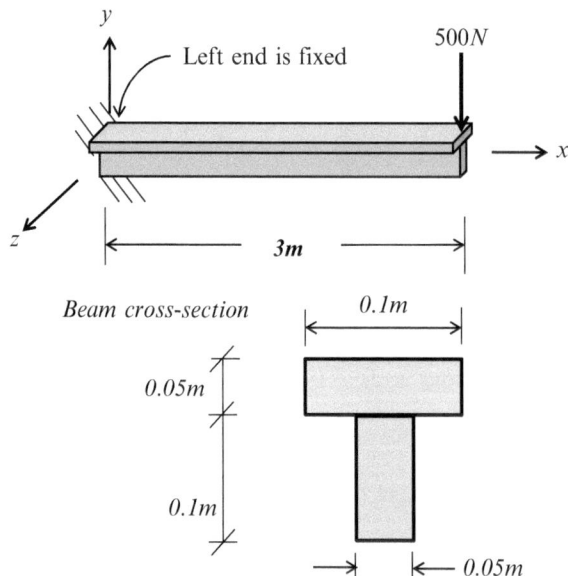

y

Left end is fixed

500N

x

z

$3m$

Beam cross-section

0.1m

0.05m

0.1m

0.05m

REQUIRED: Calculate the minimum number of nails required for the two pieces to perform as a single unit without sliding at the interface.

References

Allen D, Haisler W (1985) Introduction to aerospace structural analysis. Wiley, New York

Euler L (1744) Method inveniendi lineas curvas. Opera Omnia, St. Petersburg

Greenberg M (1978) Foundations of applied mathematics. Prentice-Hall, New Jersey

Oden J, Ripperger E (1981) Mechanics of elastic structures. McGraw-Hill, New York

Popov E (1998) Engineering mechanics of solids, Second edn. Prentice-Hall, New Jersey

Roark R, Young W, Budynas R, Sadegh A (1975) Formulas for stress and strain, Eighth edn. McGraw-Hill, New York

Wempner G (1995) Mechanics of solids. International Thomson Pub., Boston

Chapter 6
Stress and Failure Analysis

6.1 Introduction

Have you ever wondered why the Eiffel Tower, shown in Fig. 6.1, is shaped the way it is? As shown in the figure, there is clearly a relationship between the shape of the moment diagram for an evenly distributed loading (in this case—caused by wind loading) applied to a cantilever beam and the shape of the tower. Another example is the Firth of Forth Bridge, as shown in Fig. 6.2. Shown below the photo of the bridge is the moment diagram for an evenly distributed loading applied to a double cantilever beam (see Example Problem 5.3).

There are reasons for the striking similarities between the internal load diagrams obtained using the theoretical models developed in the previous three chapters and the shapes of the resulting structures, and while these models were initially developed by the scientific community, it was engineers like Gustav Eiffel who began applying these models in the late nineteenth century to the design of structures against failure. In the process, they also realized that structures could be designed not only to avoid failure but also to simultaneously create landmarks that are both visually appealing and cost effective. Structures such as the Eiffel Tower, completed in 1889, the Firth of Forth Bridge, completed in 1890, and the Brooklyn Bridge, completed in 1883 and shown in Fig. 6.3, are among the first significant structures built on earth that utilized modern mechanics to produce visually stunning structures that are nonetheless structurally sound. After these masterpieces were completed, it was not long before other similar structures began appearing all over the world, as shown in Fig. 6.4.

The twentieth century produced a plethora of new design methodologies for structures that are based on the mechanics models introduced in this course. The interested observer need only look as far as automobiles, bridges, buildings, aircraft, spacecraft, and modern windmills to see the worldwide impact of these models. Indeed, the design of essentially all modern load carrying structures emanates from the concepts developed in the previous three chapters.

D.H. Allen, *Introduction to the Mechanics of Deformable Solids: Bars and Beams,*
DOI 10.1007/978-1-4614-4003-1_6, © Springer Science+Business Media New York 2013

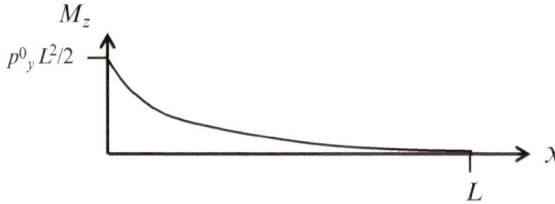

Fig. 6.1 Photo of the Eiffel Tower rotated 90° and depicted above the moment diagram for a cantilever beam with evenly distributed loading

A successful structural design requires that the object satisfy all of the design constraints. If the design fails to satisfy even one of the design constraints it has failed. One of the most famous structural failures in modern times is the case of the ill-fated RMS Titanic (Fig. 6.5). As the reader may well know, at the time the ship was built (1912) it was the largest passenger liner in the world. It went down in the North Atlantic on its maiden voyage when it struck an iceberg. There were 1,517 persons killed in the disaster.

Investigations determined that the ship went down due to a complex series of related design flaws. Large ships are designed with large bulkheads along their length so that if the hull is pierced, several of the bulkhead-separated compartments can flood without causing the ship to sink. In the case of the Titanic, the ship was designed to remain afloat if the first four bulkheads were flooded. Unfortunately, when the ship struck the iceberg, it scraped along the starboard (right) side for nearly the entire length of the ship. The ship had exposed rivets along its entire length, and these rivets were made of a somewhat brittle steel, that may have been further embrittled by the cold waters in the North Atlantic. At any rate, the iceberg sheared off the heads of these rivets, allowing water to begin flooding the first five compartments. Furthermore, as the ship canted from the flooding in the compartments, water poured over the tops of some of the bulkheads (another design flaw), causing the bulkheads aft to flood more rapidly. The Titanic disaster caused a world-wide calamity that led to significant changes in the design of modern ships.

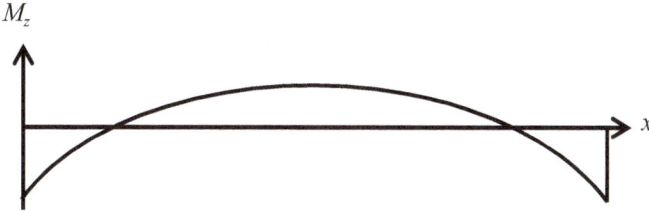

Fig. 6.2 Photo of the Firth of Forth bridge depicted above the moment diagram for a double cantilever beam with evenly distributed loading

Fig. 6.3 Photo of the Brooklyn Bridge

Fig. 6.4 Photos of bridges: Sydney Harbor Bridge (*top*); Golden Gate Bridge (*center*); and the bridge at Brazilia (*bottom*)

Although the design constraints are often known precisely, this is not always the case. An example of a case wherein the design constraints were not known a priori sufficiently to avert structural failure is the 1986 failure of the Space Shuttle Challenger, shown in Fig. 6.6. In that case, the subfreezing temperatures on the morning of the launch later proved to be the cause of the failure of the O-rings in the shuttle rocket motor booster casings. Indeed, the subject of failure prediction is still very much an open issue in the fields of solid and structural mechanics today. Due to the advanced nature of the physics involved, much of the subject of failure falls outside the scope of this text. Thus, for the sake of simplicity, we will concern ourselves herein with relatively simple and straightforward failure models.

Fig. 6.5   Photo of RMS Titanic departing Southampton on April 10, 1912, photo by F. G. O. Stuart

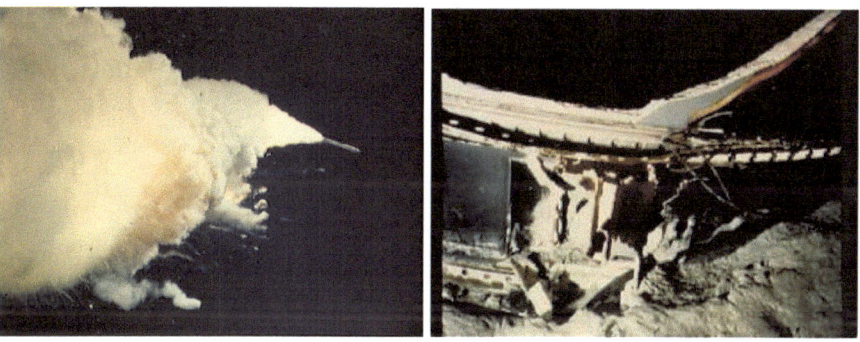

Fig. 6.6   Photos of the Challenger disaster; explosion on the *left*; Challenger underwater on the *right* (photos courtesy NASA)

6.2   Modes of Failure in Solids

6.2.1   Nonphysical Failure

Solids can fail by physical or nonphysical means. By nonphysical means, it is implied that the solid continues to perform its intended task physically, but it has nevertheless failed to meet expectations. The most common nonphysical failure modes are (1) excessive cost and (2) aesthetic failure. The former case is quite obvious; an example is the Channel tunnel (called the Chunnel) underneath the English Channel,

Fig. 6.7 Photo of full scale model of section of Chunnel at National Railway Museum in York, England

connecting England and France, as shown in Fig. 6.7. The Chunnel has failed because the enormous cost caused the debt to be unrecoverable in the expected time span due to the rise of low-cost alternative means of travel such as air. The Chunnel has not failed physically, but it has nonetheless failed to perform its intended task. An example of an aesthetic failure is the Charles De Gaulle Airport Terminal 1, shown in Fig. 6.8. While this terminal was thought to be ahead of its time when it was first opened some 40 years ago, it has become inefficient and unsightly in recent times.

6.2.2   Nonmechanical Physical Failure

There are numerous possible modes of physical failure in solids. Broadly speaking, failure can be induced chemically, thermally, or even electromagnetically. A simple example of a thermally induced failure would be melting of a solid. For example, the failure of the Space Shuttle Columbia in 2003 seems to have been induced at least in part due to overheating during reentry to the Earth's atmosphere. Another example,

Fig. 6.8 Aerial photo of Charles De Gaulle Airport Terminal 1

Fig. 6.9 Photo of Aloha Airlines disaster, photo courtesy FAA

that of chemical failure, occurs when corrosion reduces the load carrying capability of a structure, such as in the case of the 1988 Aloha Airlines disaster, shown in Fig. 6.9, where corrosion due to the salty environment in the Hawaiian Islands contributed to the development of fatigue cracks in the aircraft fuselage. Still another example is that of electromagnetically induced fracture, which can occur in computer chips due to coupling between electromagnetic and mechanical processes.

Fig. 6.10 Photos of the Wasa: bow photo on *left*, stern photo on *right*

All of the above possibilities are known to occur in nature, and each must be considered when circumstances necessitate. However, for the current course, we will restrict ourselves to a relatively small range of failure modes in solids in order to demonstrate the design process without introducing unnecessary complexity.

6.2.3 Mechanically Induced Failure

Mechanical failure normally occurs in one of four different ways: permanent deformation, fracture, excessive deformation, or structural instability. Of these, fracture is the most complex and least understood. Excessive deformation is well understood and is usually easy to account for within the design process. Consider for example a diving board. The designer will necessarily want to ensure that the board deflects neither too little nor too much for the typical person jumping on the end of the board, and this can be accounted for by utilizing the deflection equation for a cantilever beam to size the beam appropriately. Another example is the case of an aircraft wing. The designer must ensure that the wing is stiff enough to avoid having the wing tip touch down on landing, as the resulting impact could cause serious damage and even loss of life.

Structural instability is another mechanically induced mode of failure. A rather simple example of instability (although not structural) is the case of the ill-fated Wasa, a Swedish warship that rolled over and sank in a light breeze in the harbor in Stockholm on her maiden voyage in 1628, as shown in Fig. 6.10. This type of failure is beyond the scope of the current course. However, it remains an important failure mode to be considered, so that the interested student will want to delve further into the subject should this possibility exist for the structure under consideration.

Fig. 6.11 Halfdome at Yosemite National Park, induced by a really big granite crack

6.2.4 Failure by Fracture and/or Permanent Deformation

Solids are distinguished from fluids by the fact that they can undergo fracture, and this process can often (though not always) lead to failure of a structure to perform its intended task. Cracks can occur on very large length scales, such as Halfdome, shown in Fig. 6.11.

Small cracks can also lead to catastrophic failure, as in the case of the Sioux City UA flight 232 (DC-10) aircraft crash on July 19, 1989. It was found that cracks in the number 2 (tail mounted) engine stage 1 fan disk propagated in an unstable manner, thereby causing portions of the engine to break off and damage the aircraft tail and controls (NTSB 1990). The aircraft subsequently broke up during an emergency landing at Sioux Gateway Airport. Although 185 passengers survived, 111 passengers were killed. The fan disk was found in a cornfield 3 months after the disaster and it was reconstructed. A photograph of the reconstructed fan disk, showing the cracks in the fan disk, is shown in Fig. 6.12 (NTSB 1990).

Another recent major failure was the collapse of the I-35 W Mississippi River Bridge in Minneapolis on August 1, 2007, shown in Fig. 6.13. A postmortem inspection of the bridge revealed that the beam connection plates had corroded over time, thus reducing the material properties of the plates. This corrosion, together with long-term bridge overloading caused by adding two lanes of traffic to the initial design, contributed to unstable crack propagation and collapse of several sections of the bridge. Thirteen people were killed and 145 people were injured.

Fig. 6.12 Photo of reconstruction of the stage 1 fan disk in the Sioux City aircraft crash of flight UA 232 on July 19, 1989 (NTSB 1990), photo courtesy NTSB

Fig. 6.13 Photo on *left* of the Minneapolis Bridge Collapse (courtesy US Coast Guard); Photo on *right* showing fracture in gusset plate (photo courtesy NTSB, 1990)

The ability to predict when a crack will grow and where it will go depends on the material utilized in the solid under consideration. Brittle solids generally are the class of materials for which fracture can be most easily predicted. However, in the case where the brittle solid of interest is not isotropic, such as a laminated continuous carbon fiber composite currently being deployed in the Boeing Dreamliner and the Airbus A380, shown in Fig. 6.14, the prediction of fracture is

Fig. 6.14 Boeing 787 Dreamliner on *left*; Airbus A380 on *right*

an advanced topic that is beyond the scope of the current text. Also, predicting fracture in ductile solids, especially due to cyclic loads, is a complicated subject that is left for more advanced coursework. Nevertheless, it is often possible to design solid components so as to obviate the possibility of fracture and/or permanent deformation with little more than a maximum stress criterion. The exact form of this criterion will depend on the material under consideration, and the framework for the criterion can be at least partially deduced from theoretical considerations (such as material symmetry), but will ultimately require a suite of experiments to be performed in the laboratory in order to account for the phenomenological nature of crack growth in complex materials.

In order to develop a rudimentary understanding of how structural materials fracture, consider first the case of a uniaxial bar loaded as shown in Fig. 6.15. There are several different ways that the bar may fail, but two of the most common ways are by fracture on a plane normal to the loading direction or by fracture on a plane that is rotated 45 ° with respect to the loading direction, as shown in the figure. The orientation of the failure surface generally depends on the material makeup of the bar. A careful analysis of the bar using Newton's first law will reveal that the plane normal to the loading direction is the plane of the maximum normal stress, given by

$$\sigma_{xx} = F/A$$

where F is the applied load and A is the cross-sectional area of the bar, as shown in the figure. Alternatively, the plane that is rotated at a 45 ° angle from the loading direction is the plane of the maximum shear stress in the bar, given by

$$\sigma_{x'y'} = F/2A$$

as shown in Fig. 6.16. Thus, it can be seen that a uniaxial bar may fail on the plane of maximum shear stress or the plane of maximum normal stress, and the plane on which this occurs depends on the type of material being tested.

The example shown in Figs. 6.15 and 6.16 thus demonstrates two different possible fracture modes when a bar is loaded uniaxially to failure. For components

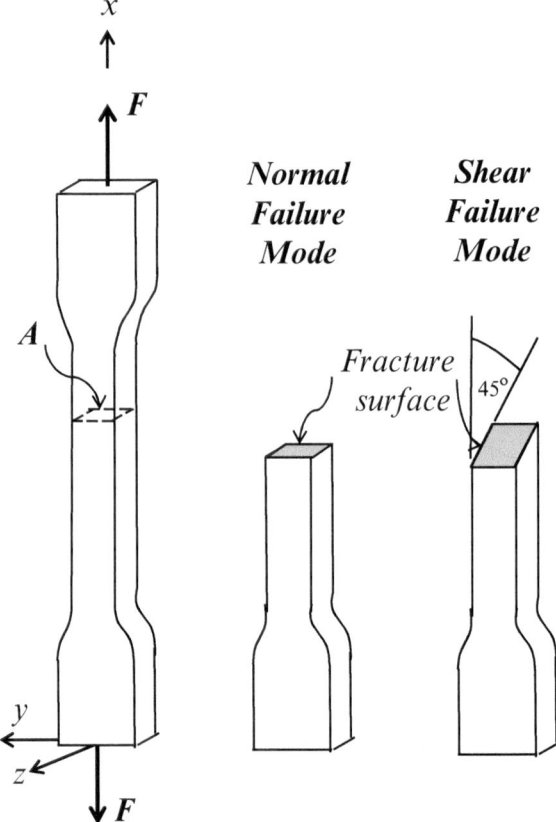

Fig. 6.15 A typical uniaxial bar test showing failure modes

subjected to more complex loading conditions, there are generally three different modes of crack propagation, as shown in Fig. 6.17.

It can be concluded from the above discussion that fracture is related to the stress state on the failure planes in solids that are loaded uniaxially. This same conclusion applies to solid objects subjected to loadings that produce more complicated stress states. However, in solids subjected to more complicated loadings, it is not generally known at what points in the object and on what planes the stress components are critical a priori. Therefore, the practicing structural analyst will find it necessary to select a set of coordinate axes arbitrarily and to utilize these axes to predict the state of stress at every point in the object using this arbitrarily chosen set of axes. Once the stress components are predicted in this arbitrarily chosen set of coordinates, it will then be necessary to perform a further analysis to determine at what point in the object and on what orientation the components of stress are sufficiently elevated to initiate fracture. This part of the analysis requires the ability to perform stress transformations, to be discussed in the next section.

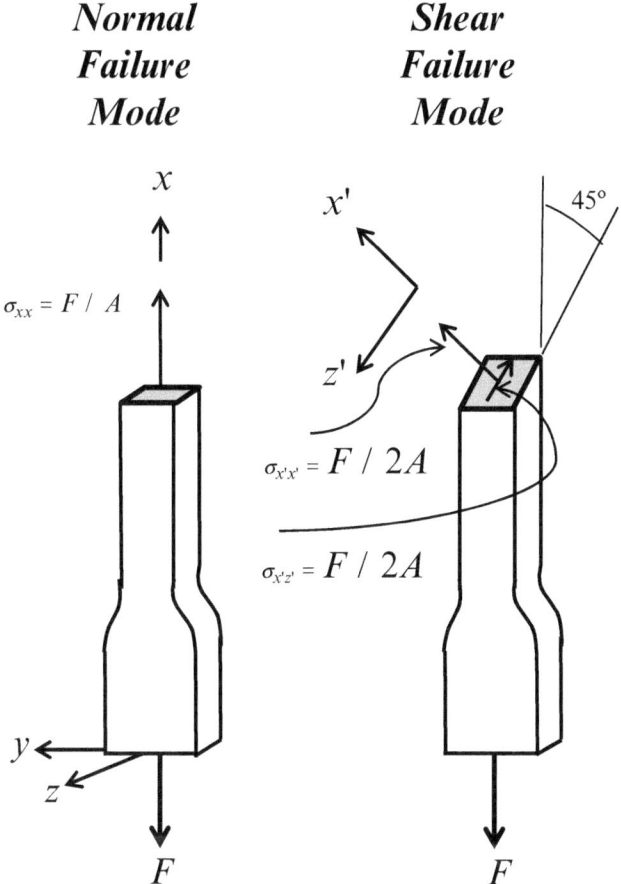

Fig. 6.16 Resultant stresses a uniaxial bar loaded to failure in two different modes

6.3   Stress Transformations

Augustin Cauchy introduced the modern interpretation of stress in 1822. In his paper, Cauchy proved that the nine components of stress on the three faces normal to the three coordinate axes are sufficient to uniquely define the loading (or kinetic) state at any point in a body, and this proof is called Cauchy's formula. This study of stress made by Cauchy has led to much detailed examination of the physical nature of stress over the past two centuries. Since the coordinate axes that are chosen for the purpose of analyzing a given body are entirely arbitrary, meaning that absolutely **any** chosen set of Cartesian coordinates is physically acceptable, then it follows that two different observers could perform an analysis of a given object with two different sets of Cartesian coordinates, and if both analyses are performed

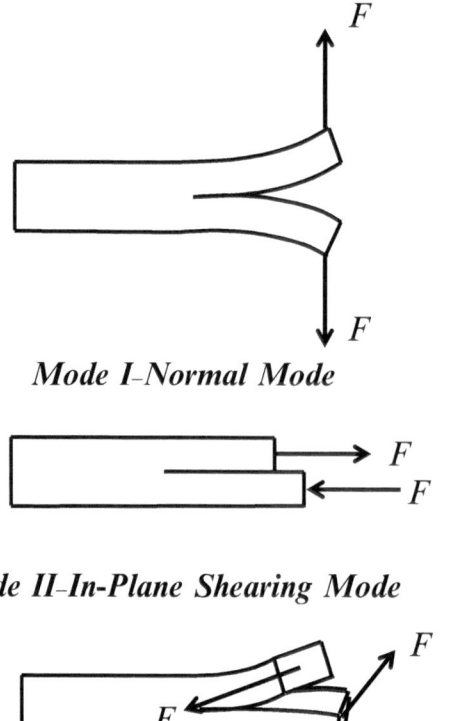

Mode I–Normal Mode

Mode II–In-Plane Shearing Mode

Mode III–Out-of-Plane Shearing Mode

Fig. 6.17 Fracture modes in solids

using the same model and are mathematically correct, then it should be possible to compare the stresses obtained in the two analyses for the purpose of showing that they are physically equivalent, as demonstrated in Fig. 6.18.

Since the coordinate locations of these two sets of coordinate axes can be related trivially, we will not concern ourselves with this part. On the other hand, the **orientations** of these two sets of coordinate axes produce complicated relations between the stresses in each coordinate system, so that relating the stress predictions obtained by the two different observers is not a trivial matter by any means. *Our aim is to do just that—to relate the stresses at a given point in an object with respect to two different coordinate systems that are rotated with respect to one another.*

In the structural components introduced in the current course, it is almost always the case that the significant components of stress lie in a plane, called plane stress, as shown in Fig. 6.19a. If there is a component of normal stress perpendicular to the plane described in Fig. 6.19b, then the state of stress is called generalized plane stress, and for purposes of simplicity, we have defined that plane to be the x–y plane in the current discussion. For the case of generalized plane stress, there is a

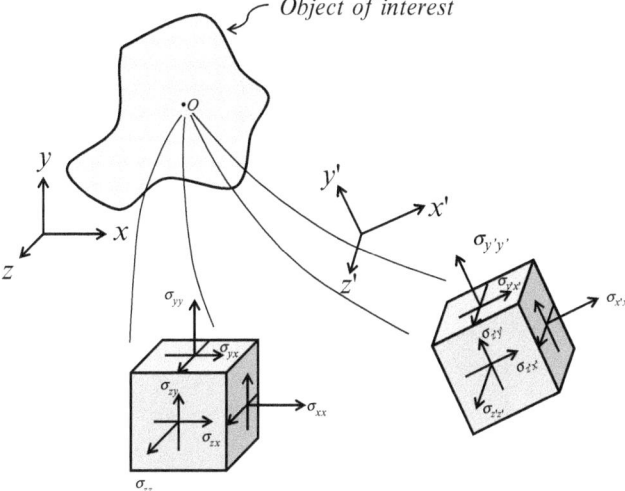

Fig. 6.18 Stress state at point O in an object analyzed with two different coordinate systems

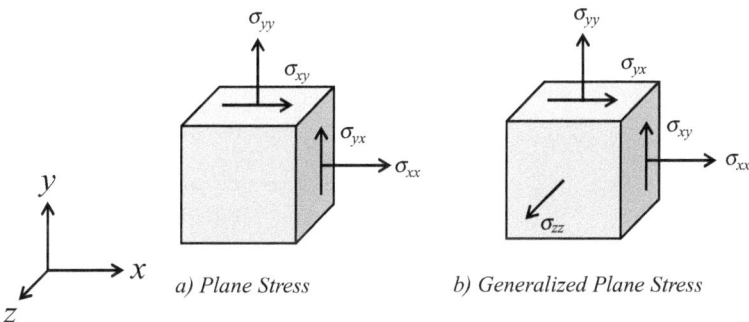

Fig. 6.19 Special cases of stress states

simplified technique for performing transformations of stress from one coordinate system to another that is rotated about the z-axis.

Now, suppose that we want to determine if the state of stress in the primed coordinate system is equivalent to the stress in the unprimed coordinate system for the case of generalized plane stress (note that the z coordinate direction is the same in both depictions), as shown in Fig. 6.20. In order to determine the relation between the stress components in the primed and unprimed coordinate systems, first pass a cutting plane through the stress block in the unprimed coordinate system rotated an angle θ about the z-axis, as shown in Fig. 6.21.

Using the free body diagram resulting from this cutting plane, as shown in Fig. 6.22, we will now sum forces in the x' coordinate direction, recalling that the stresses must be multiplied by the area over which they act in order to produce

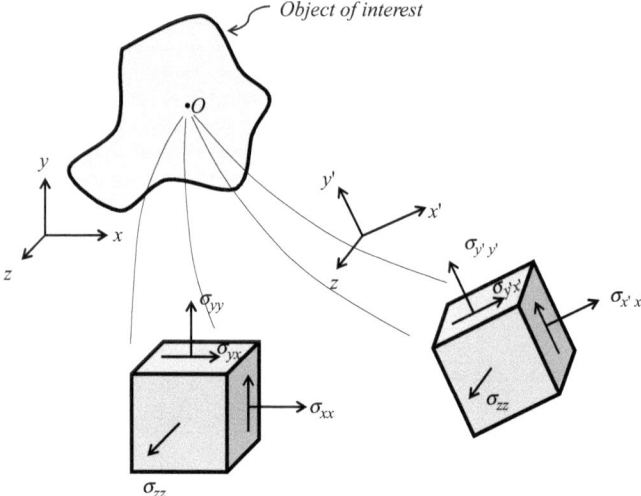

Fig. 6.20 Planar stress transformations

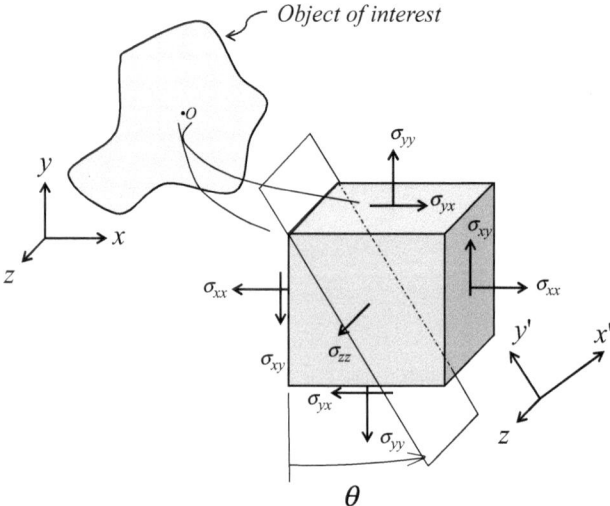

Fig. 6.21 Cutting plane in primed coordinate system passed through the stress block in the unprimed coordinate system

forces. In order to do this, the area of the face normal to the x'-axis is denoted as A. This results in the following (since $\sigma_{xy} = \sigma_{yx}$):

$$\sum F_{x'} = 0 = \sigma_{x'x'}A - \sigma_{xx}\cos\theta(A\cos\theta) - \sigma_{yy}\sin\theta(A\sin\theta)$$
$$-\sigma_{xy}\sin\theta(A\cos\theta) - \sigma_{xy}\cos\theta(A\sin\theta)$$

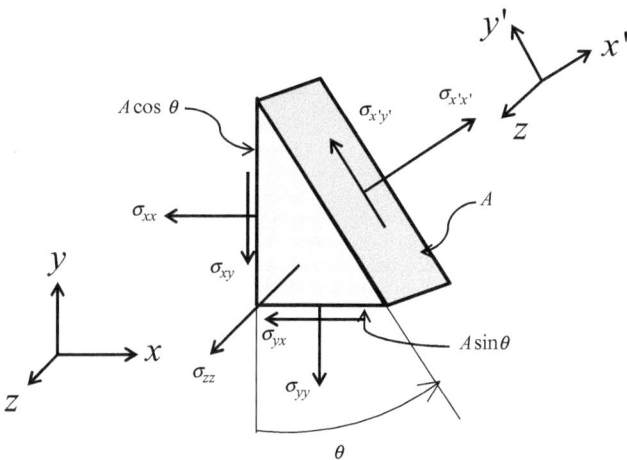

Fig. 6.22 Free body diagram of stress block with cutting plane

Rearranging terms and dividing through by A results in

$$\sigma_{x'x'} = \sigma_{xx}\cos^2\theta + \sigma_{yy}\sin^2\theta + 2\sigma_{xy}\sin\theta\,\cos\theta \qquad (6.1)$$

Similarly, summing forces in the y' direction will result in

$$\sigma_{x'y'} = -(\sigma_{xx} - \sigma_{yy})\sin\theta\cos\theta + \sigma_{xy}(\cos^2\theta - \sin^2\theta) \qquad (6.2)$$

Equations (6.1) and (6.2) can be used to calculate the components of stress in the primed coordinate system as functions of the stress components in the unprimed coordinate system for any given value of θ.

6.3.1 Mohr's Circles for Performing Coordinate Transformations of Stress

Equations (6.1) and (6.2) are cumbersome to use for analysis purposes. A more useful graphical technique was introduced by Culmann (1866). This technique was later expanded and used to great effect for the prediction of failure by Mohr (1868), from whence comes the name applied to this technique—"Mohr's circle." In order to construct this graphical technique, first recall the following trigonometric identities

$$\cos^2\theta = \frac{1 + \cos 2\theta}{2} \quad \sin^2\theta = \frac{1 - \cos 2\theta}{2} \quad 2\sin\theta\cos\theta = \sin 2\theta \qquad (6.3)$$

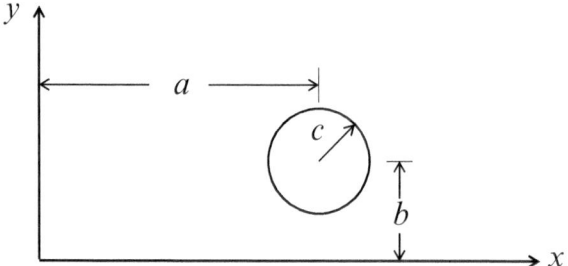

Fig. 6.23 General equation of a circle

Substituting the above identities into (6.1) and (6.2) and rearranging will result in the following two equations

$$\sigma_{x'x'} - \left(\frac{\sigma_{xx} + \sigma_{yy}}{2}\right) = \left(\frac{\sigma_{xx} - \sigma_{yy}}{2}\right) \cos 2\theta + \sigma_{xy} \sin 2\theta \tag{6.4a}$$

$$\sigma_{x'y'} = -\left(\frac{\sigma_{xx} - \sigma_{yy}}{2}\right) \sin 2\theta + \sigma_{xy} \cos 2\theta \tag{6.4b}$$

Squaring and adding the above two equations will result in the following equation

$$\left[\sigma_{x'x'} - \left(\frac{\sigma_{xx} + \sigma_{yy}}{2}\right)\right]^2 + [\sigma_{x'y'} - 0]^2 \left[\sqrt{\left(\frac{\sigma_{xx} - \sigma_{yy}}{2}\right)^2 + \sigma_{xy}^2}\right]^2 \tag{6.5}$$

It can be seen that the above equation is equivalent to the general equation of a circle, shown in Fig. 6.23, and given by

$$[x - a]^2 + [y - b]^2 = [c]^2 \tag{6.6}$$

Where, by comparison of (6.5) and (6.6), the following transformation of variables is apparent

$$x \rightarrow \sigma_{x'x'}$$
$$y \rightarrow \sigma_{x'y'}$$
$$a \rightarrow \frac{\sigma_{xx} + \sigma_{xy}}{2}$$
$$b \rightarrow 0$$
$$c \rightarrow \sqrt{\left(\frac{\sigma_{xx} - \sigma_{yy}}{2}\right)^2 + \sigma_{xy}^2} \tag{6.7}$$

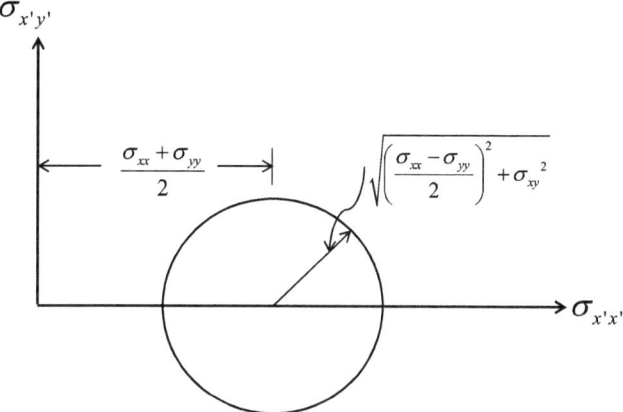

Fig. 6.24 Mohr's circle for plane stress

Applying the above transformations to the circle shown in Fig. 6.23 results in the graphical representation shown in Fig. 6.24.

Note that all possible orientations for a plane stress state at a given material point in an object that are rotated about the z-axis lie on the circle shown in Fig. 6.24. Furthermore, because we have employed the trigonometric half angle formulas in (6.3), angles shown in Fig. 6.24 are exactly twice as large as angles depicted in the real world.

It can be shown that for the case of generalized plane stress (σ_{xx}, σ_{xy}, σ_{yy}, $\sigma_{zz} \neq 0$, $\sigma_{yz} = \sigma_{xz} = 0$), the depiction of Mohr's circle shown in Fig. 6.24 can be generalized to produce three circles, as shown in Fig. 6.25.

Mohr's circle was created nearly a century and a half ago, in a time when graphical techniques were the most powerful mathematical tools for many engineering applications. An early example (in 1867) of the physical power of this method is demonstrated in Fig. 6.26 (Culmann 1866; Meyer 1867; Wolff 1870; Thompson 1917). The figure shows the lines of principal stresses in a crane analyzed by Culmann on the left. On the right is a depiction by Wolff of the trabecular alignment in the proximal femur of a human. This impressive demonstration of the importance of principal stresses is said to have occurred by coincidence when Professor Culmann visited the dissecting room of his colleague Hermann Meyer in Zurich. Upon seeing a section of bone, Culmann is said to have cried out, "That's my crane!" (Thompson 1917).

Today it is no longer necessary to utilize such antiquated methods due to the power of computers. Nonetheless, there is much information to be extracted from a careful study of Mohr's circles that is physically significant with respect to failure of solids. Several important deductions are given in Table 6.1.

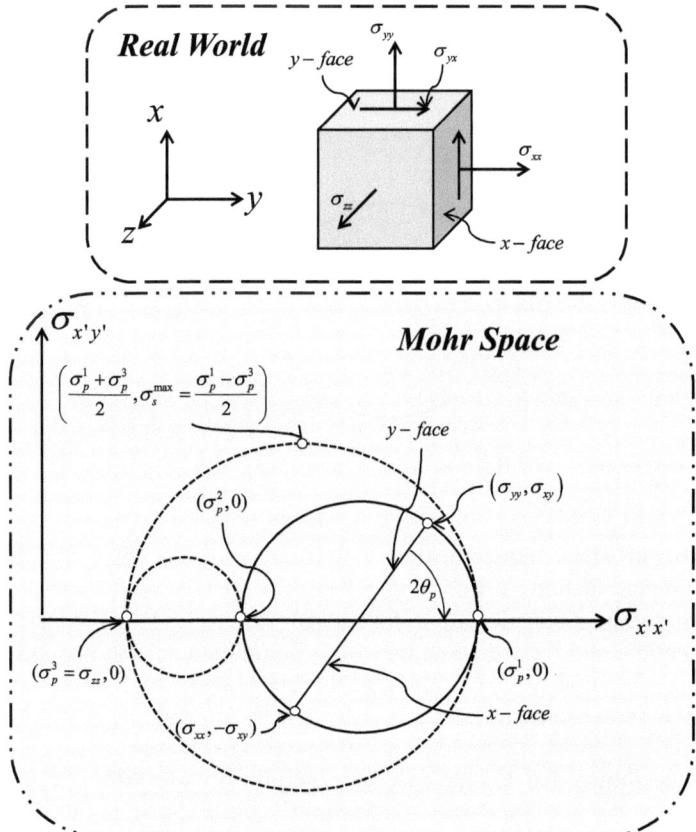

Fig. 6.25 Mohr's circles for generalized plane stress

There are several canned programs available on the internet at no cost to the user. By importing the values of the stress components for a given material point in an object with respect to an arbitrary set of coordinate axes (remember—for generalized plane stress only!), the user can have Mohr's circles plotted by the software. In addition, there are a few canned software programs that will even construct diagrams of the material point, with the stress components labeled on the diagrams. It is therefore far less difficult to perform analysis of stress at a given material point than in former times, when computer algorithms were not available for this purpose.

For those who prefer to take the time to construct their own Mohr's circles graphically, Table 6.2 presents the procedure for doing so.

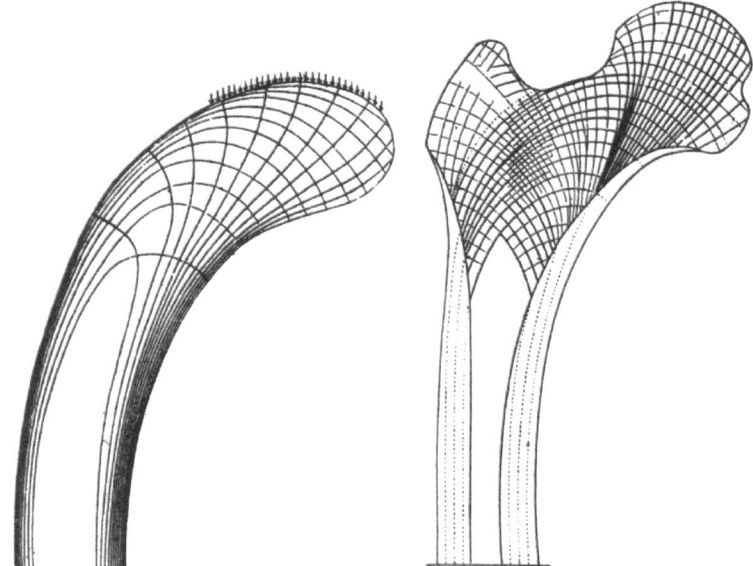

Fig. 6.26 Lines of principal planes (*left*) in a curved crane compared to the trabecular alignment in a human proximal femur (*right*) (Culmann 1866; Meyer 1867; Wolff 1870; Thompson 1917)

Table 6.1 Properties of stress deduced from Mohr's circles

Important deductions from Mohr's circles

- The three circles subtend the horizontal axis at points labeled σ_p^1, σ_p^2, and σ_p^3. These normal stresses are called principal stresses because the shear stresses on the planes are zero.
- A vertical diameter of the largest circle subtends the circle at the top and bottom, where the shear stress attains it maximum, σ_s^{max}.
- The angle (in Mohr space) between the x- (or y-) face and a horizontal diameter of the circle is twice the angle from the x- (or y-) face to a principal plane in the real world (denoted as $2\theta_p$ on Mohr's circle).
- Since principal planes and planes of maximum shear stress are always normal to one another in Mohr space, they are exactly 45 ° apart in the real world.
- For the case wherein $\sigma_p^1 \geq \sigma_p^2 \geq \sigma_p^3$, it follows that $\sigma_s^{max} = (\sigma_p^1 - \sigma_p^3)/2$ is the diameter of the largest circle.

Table 6.2 Procedure for drawing Mohr's circles

STEP 1: Plot the coordinates of the x-face $(\sigma_{xx}, \pm\sigma_{xy})$ and the y-face $(\sigma_{yy}, \mp\sigma_{xy})$ using the sign convention that *shear stresses that cause a clockwise couple are positive, and shear stresses that cause a counterclockwise couple are negative.*

STEP 2: Draw a straight line connecting these two points.

STEP 3: The intersection of the line drawn in step 2 and the horizontal axis is the center of the circle. Use a compass to draw the circle that passes through the two points drawn in step 1.

STEP 4: Label the diameter of the circle that ends at the two points plotted in step 1 as the x-face and y-face as appropriate.

STEP 5: Label the three principal stresses, including the out-of-plane normal stress (even if it is zero!), and draw the two remaining circles, as depicted in Fig. 6.24.

STEP 6: Label the maximum shear stress at the top of the largest circle, σ_s^{max}, and calculate the value of the angle $2\theta_p$, also depicted in Fig. 6.24.

Example Problem 6.1

Given: Suppose that one or more of the models developed in this text has been used to predict the state of stress as a function of location in a bar with known loads, geometry, and materials properties, and this analysis has resulted in the stress state shown below at a point in the object identified by the designer as a critical point where failure of the object may be initiated.

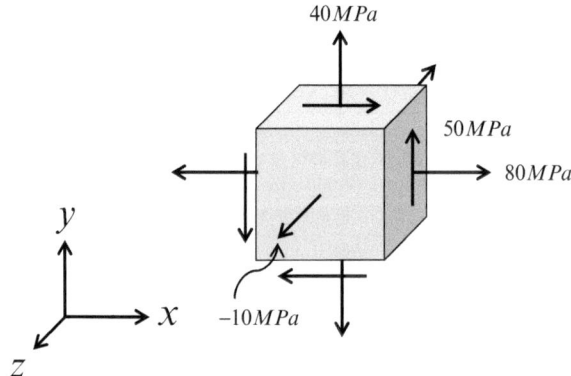

Required

(1) Plot Mohr's circles and determine the principal stresses and the maximum shear stress.

(2) Draw sketches of the principal planes and the plane of maximum shear stress, showing the stress components on these planes.

Solution

1. Mohr's circles.

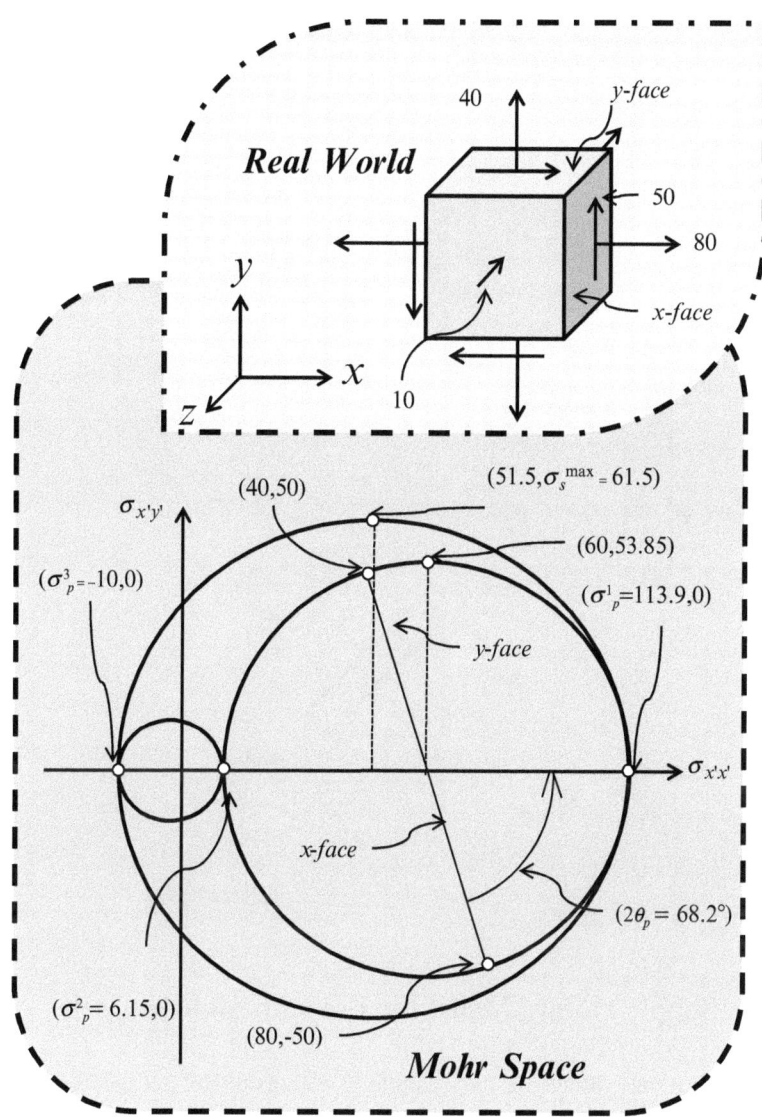

2. Sketches in the real world.

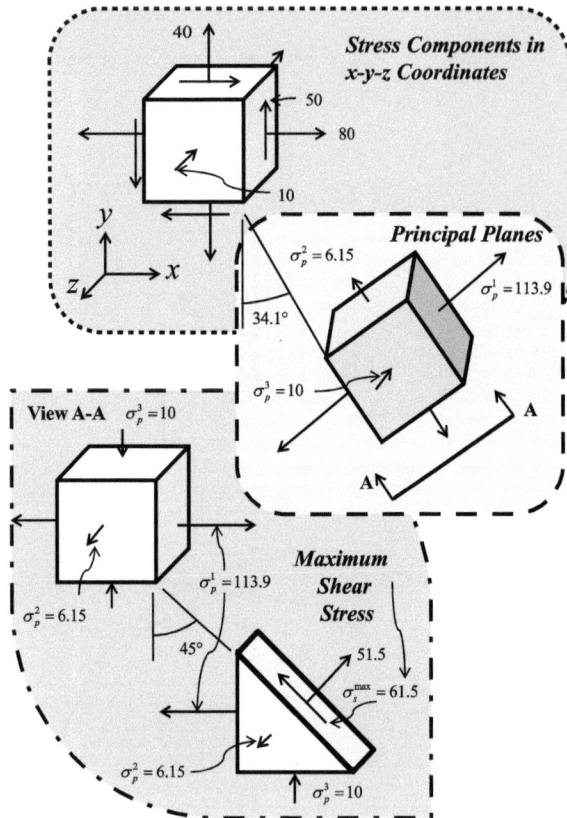

6.4   A Simple Failure Model for Use with Mohr's Circle

In Sect. 6.2, it was shown via the example of a uniaxial bar test that the plane on which the maximum normal stress and the maximum shear stress occur correspond to the planes on which fracture occurs, depending on whether the fracture event is mode I (normal mode) or mode II or III (shearing modes). This is often the case even for more complicated stress states, and indeed the values of the stresses obtained from the uniaxial test at failure can be utilized as predictors of failure due to fracture for more complicated stress states. Prediction of failure due to excessive normal or shear stress is easily accommodated by employing Mohr's circle.

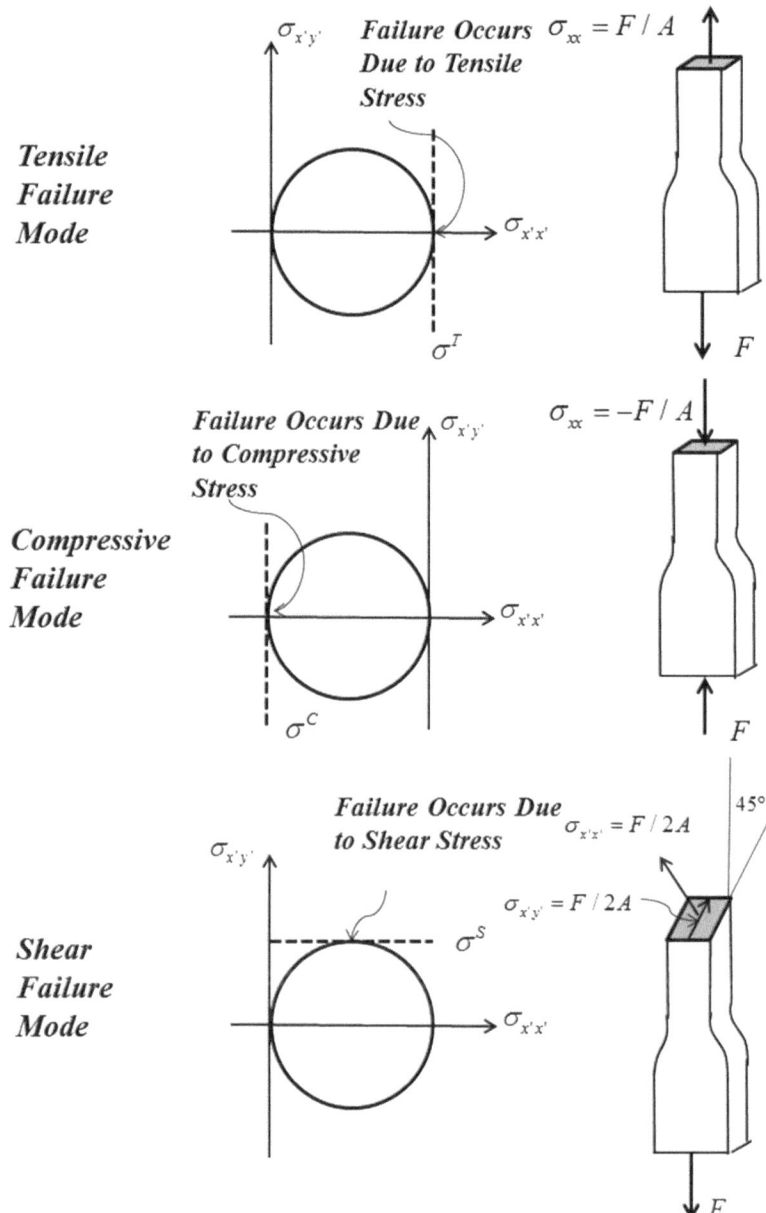

Fig. 6.27 Fracture modes in uniaxial bars

For example, in some cases fracture and/or yielding is initiated in materials when the tensile normal stress reaches a critical value, termed σ^T. Similarly, fracture and/or yielding may be initiated when the maximum compressive stress reaches a critical value, termed σ^C. Furthermore, failure due to fracture and/or yielding may

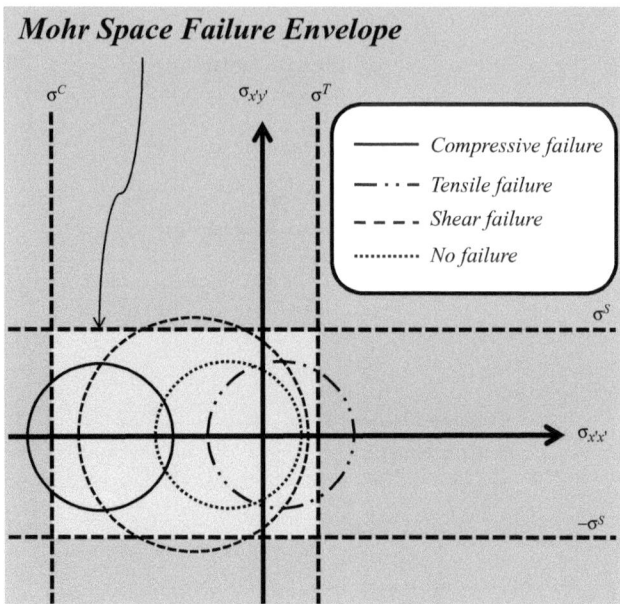

Fig. 6.28 A simple failure envelope based on maximum tensile, compressive, and shear stress

be initiated in a material when the shear stress exceeds a critical value, termed σ^S. These three uniaxial tests and their corresponding Mohr's circles are depicted in Fig. 6.27. The values obtained from these tests can then be utilized to construct a *failure envelope* for purposes of predicting failure of solids subjected to more complicated stress states, as depicted in Fig. 6.28. Thus, it is possible to use Mohr's circle as a means of designing a solid to avoid failure by yielding and/or fracture.

While the approach described in Fig. 6.28 is a bit simplistic (and also inaccurate!), a number of more advanced failure models have been developed for a variety of materials. These seem to have been first studied by Charles Coulomb for geologic media in the late eighteenth century. It is a testament to the power of his models for friction that many of his theories are still in practice today. In the nineteenth century perhaps the first model for predicting yielding in isotropic metals is due to Henri Tresca. Later models were developed by Richard Von Mises and others, including Otto Mohr. Early in the twentieth century research turned towards the development of models for predicting fracture in solids beginning with the work of A.A. Griffith, and this work continues at the time of this writing. It must be stated that the prediction of failure due to yielding and/or fracture is an advanced subject, and as such is beyond the scope of the current text. However, for purposes of the current text suffice it to say this—*if a model is capable of accurately predicting the state of stress at every point in a solid object, then failure models can be employed to make predictions as to whether or not the object is likely to fail due to fracture and/or yielding.* This approach to modeling failure can be stated mathematically as follows. Given a material property, $F = F(\sigma_{xx}, \sigma_{yy}, \sigma_{zz}, \sigma_{yz}, \sigma_{xz}, \sigma_{xy})$, then for a given state of stress at a point in a solid

$$F < 0 \Rightarrow failure\ does\ not\ occur$$

$$F \geq 0 \Rightarrow \ failure\ occurs \qquad\qquad (6.8)$$

In this text only a simple (and oftentimes inaccurate!) model has been employed for the purpose of demonstrating how failure may be predicted due to fracture and/ or yielding. It remains for the student who is interested in this subject to delve further into recent research that is available in the open literature on the subject of the prediction of failure of solids due to yielding and/or fracture.

Example Problem 6.2
Given: An analysis of a structural component results in the state of stress shown below at a point where the designer is concerned that yielding may occur.

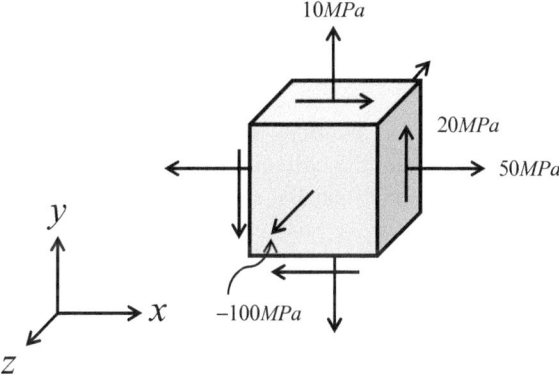

$$\sigma^{T} = 100\ \text{MPa}, \quad \sigma^{C} = 200\ \text{MPa}, \quad \sigma^{S} = 50\ \text{MPa}$$

Required

1. Check for yielding using Mohr's circle.
2. If yielding occurs, determine the maximum allowable compressive stress in the z direction in order to avoid yielding.

Solution

1. Mohr's circles.

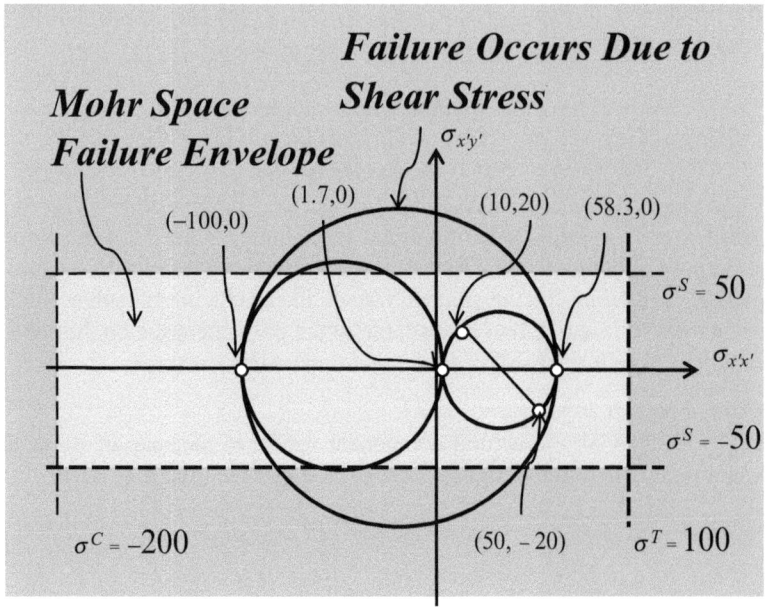

2. Since yielding occurs due to shearing, the radius of the largest circle cannot exceed $\sigma^S = 50\,\text{MPa}$. Therefore, the maximum allowable value of the third principal stress, σ_{zz}^{max}, can be calculated as follows

$$\frac{(-\sigma_{zz}^{max} + 58.3)}{2} = 50 \Rightarrow \sigma_{zz}^{max} = \boxed{-20.85\,\text{MPa}} \quad \text{Q.E.D.}$$

6.5   Assignments

PROBLEM 6.1

GIVEN: An engineer has performed a stress analysis of the object below using the unprimed coordinate axes shown. That analysis has produced the state of stress shown in the diagram at Point O. All stress components are in MPa.

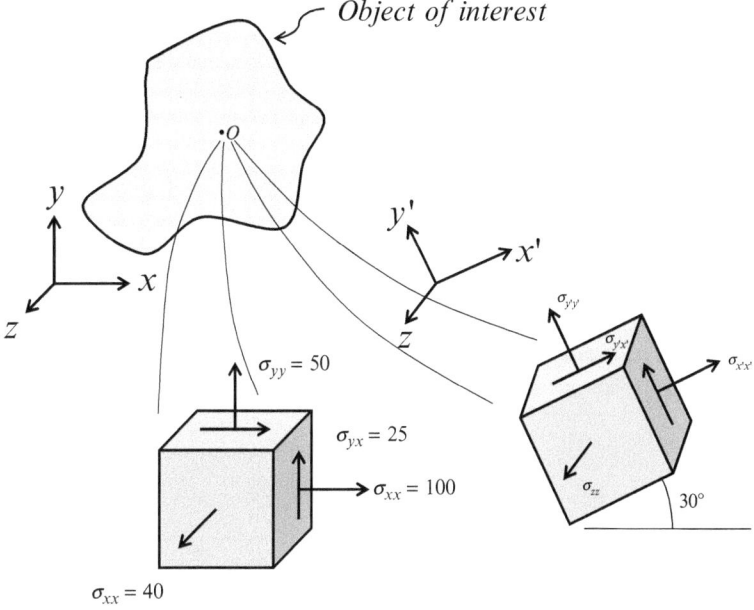

REQUIRED

1. Calculate the stress components $\sigma_{x'x'}$, $\sigma_{x'y'}$, $\sigma_{y'y'}$ in the primed coordinate system assuming that the primed coordinate system is rotated 30 ° counterclockwise about the z-axis as shown in the diagram.

PROBLEM 6.2
GIVEN: From Problem 6.1, the state of stress at Point O in an object is as shown below. All stress components are in MPa.

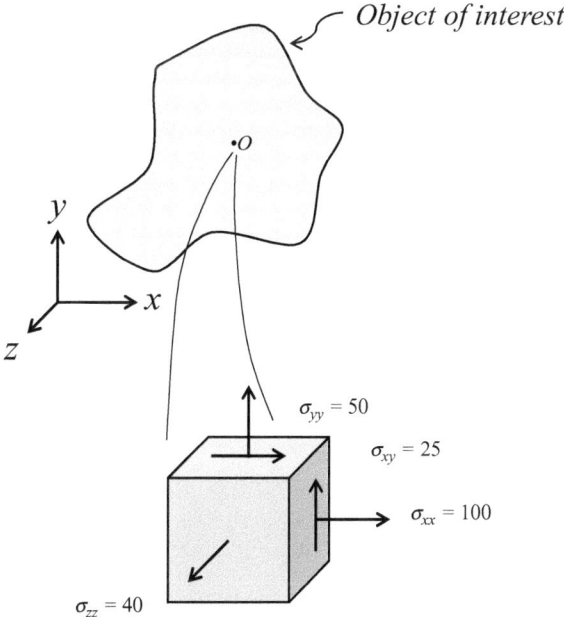

REQUIRED

1. Draw the three Mohr's circles at Point O on a piece of graph paper using a compass.
2. Label the points $(\sigma_{xx}, \sigma_{xy})$, $(\sigma_{yy}, \sigma_{xy})$, and $(\sigma_{zz}, 0)$ on the graph.
3. Label the x- and y-planes on the circle.
4. Label the three principal stresses on the graph and calculate their values.
5. Label the maximum shear stress on the graph and calculate its value.
6. Label the face rotated 30 ° counterclockwise from the x-face and calculate the state of stress on this face (from Problem 6.1).

PROBLEM 6.3

GIVEN: From Problems 6.1 and 6.2, the state of stress at Point O in an object is as shown below. All stress components are in MPa.

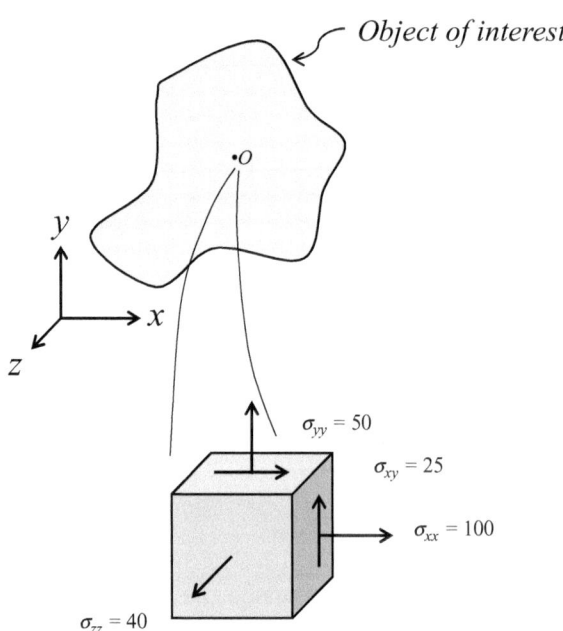

REQUIRED: Using the three Mohr's circles drawn in Problem 6.2

1. Draw sketches showing the principal planes.
2. Draw a sketch showing the plane of maximum shear stress.

PROBLEM 6.4

GIVEN: An analysis of a structural component reveals the state of stress at a point O as shown below. All stress components are in MPa. The failure stresses for the material in question are as follows

$$\sigma^{\mathrm{T}} = 250\,\mathrm{MPa},\ \sigma^{\mathrm{C}} = 350\,\mathrm{MPa},\ \sigma^{\mathrm{S}} = 100\,\mathrm{MPa}.$$

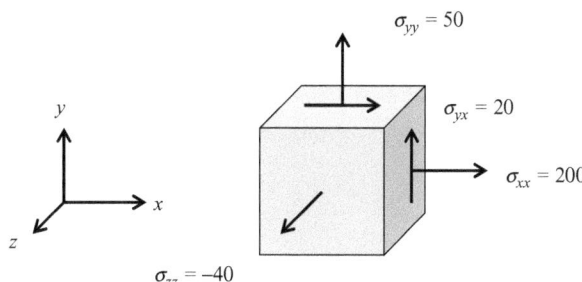

REQUIRED

1. Draw the three Mohr's circles.
2. Draw sketches showing the principal planes and the plane of maximum shear stress.
3. Plot the failure envelope on the graph showing Mohr's circles in part (1) above.
4. Determine if failure is predicted and if so tell what mode.

References

Culmann R (1866) Die Graphische Statik. Meyer & Zeller, Zurich

Meyer H (1867) Die arkitectur der spongiosa. Arch Anat Phys 47:615–28

Mohr O (1868) Z architek u ing ver. Hannover, Germany

National Transportation Safety Board, Aircraft Accident Report (1990). United Airlines Flight 232 Sioux City Gateway Airport, 19 Jul 1989 (NTSB/AAR-90/06)

Thompson DW (1917) Growth and form. Cambridge University, Cambridge

Wolff J (1870) Die innere architektur der knochen. Arch. Anat. Phys. 50

Chapter 7
Introduction to Structural Design

7.1 Introduction

We define a structural component as any object or portion of an object designed to carry mechanical loading. In most (but not all) cases, structural components are intended to perform within their linear elastic range of material behavior. Structural design is perhaps as old as mankind itself, although little evidence exists prior to about 6,000 years ago.

Possibly the most famous example of ancient structural design dates to the pyramids of Egypt. At that time design was essentially experimental in nature. The first pyramid, built by Imhotep, the first engineer known to us by name, for the pharaoh Djoser, is called the Step Pyramid because it was composed of successively smaller mastabas constructed one on top of the other, as shown in Fig. 7.1. Although this pyramid is badly decayed and somewhat simplistic, it can still be seen at Saqqara west of the Nile River today. The second, third, and fourth pyramids, built for Sneferu shortly after the step pyramid, seem to have undergone considerable development which we might today term "design," as shown in Figs. 7.2, 7.3, and 7.4, respectively. Archeologists today believe that this design process was largely experimental in nature, and in fact, the first of these pyramids, called the Meidum Pyramid, partially collapsed, while the second, called the Bent Pyramid, had to be redesigned in mid-construction in order to avoid collapse. Finally, the third pyramid built for Sneferu, called the Red Pyramid was successfully completed, and this pyramid, the second largest ever built, remains intact today.

Amazingly, all four of these pyramids were built more than 4,500 years ago, and it is estimated that the last three took approximately 30 years to complete. This may be the most monumental example of experimental design in recorded history. In fact, it has been suggested that the cost to Egypt of these mammoth construction projects was sufficient to cause economic problems throughout the country. Thus, cost as a structural design constraint is by no means new.

D.H. Allen, *Introduction to the Mechanics of Deformable Solids: Bars and Beams*,
DOI 10.1007/978-1-4614-4003-1_7, © Springer Science+Business Media New York 2013

Fig. 7.1 Photo of the step pyramid of Djoser

Fig. 7.2 Photo of Sneferu's Meidum Pyramid

The Greeks and Romans are known to have developed rules of thumb for the purposes of designing structural components, and at least some of these rules were used well into the middle ages. Perhaps the most prominent examples of experimental design during this period were the Gothic cathedrals built in the Northwest

Fig. 7.3 Photo of Sneferu's second pyramid, the Bent Pyramid

Fig. 7.4 Photo of Sneferu's third pyramid, the Red Pyramid

Fig. 7.5 Photos of the Gothic Cathedrals at Amiens (*left*) and St. Denis (*right*)

of France and Southern England. This period of fervent construction is regarded by historians to have begun around 1,140 A.D., when the cathedrals at Amiens and St. Denis were constructed with the first gothic arches, as shown in Fig. 7.5. For a period of nearly a century and a half these cathedrals were the most important construction projects in Western Europe, as clerics and their parishioners all over the region attempted to find ways to heighten the apse of the central nave of the cathedrals higher and higher. Construction began with the production of the foundation, followed by the placing of the cathedral columns. These two parts of the construction would span perhaps 20 years, during which time the cathedral was often used for services even though there was no roof. Unfortunately, in many cases, when the roof was added, the lateral loads imparted to the tops of the columns caused quite a few apses to collapse, often killing construction workers. In order to provide the necessary structural integrity to continually increase the height of the apse, experimentation revealed that secondary naves provided lateral loadings sufficient to increase the height of the central naves. When the demand for even higher apses persisted, mammoth bell towers were added to provide even more lateral strength to the cathedrals. By the end of the twelfth century, an additional experimental design innovation called flying buttresses was in use for providing further lateral structural integrity. An example of flying buttresses is shown in Fig. 7.6.

Unfortunately, as Newton's laws would later demonstrate, all structures on Earth are ultimately limited by Earth's gravitational field. Thus, despite the best efforts of engineers in the middle ages, there was a limit to the height that could be attained using stone as a building material, and this limit seems to have been reached at Beauvais, which at 48 m is the highest vault from the Gothic period, as

Fig. 7.6 Photo of flying buttresses utilized at Notre Dame cathedral in Paris

Fig. 7.7 Photo of Beauvais cathedral

shown in Fig. 7.7. Unfortunately, the cathedral partially collapsed in 1,287 A.D. and work was never completed on the cathedral. Efforts to build further cathedrals were eventually forestalled by war and plague, both of which swept across Europe in the fourteenth century. Nevertheless, the construction of Gothic cathedrals is perhaps

Fig. 7.8 Artist's rendition of the National Aerospace Plane (courtesy NASA)

the most ambitious example of experimental design on a grand scale, as nearly a hundred Gothic cathedrals were built during this period in Western Europe.

As we have seen in Chap. 1, rigorous scientifically based models were not developed until after the publication of Newton's *Principia*, which opened the door via the construction of conservation laws and calculus. More than a century later, fully three-dimensional models for deformable bodies paved the way for an explosion of design techniques in the second half of the nineteenth and first half of the twentieth centuries. Whether these developments were caused by the industrial age or vice versa is neither here nor there. The fact is that by the turn of the twentieth century rigorous structural design was well underway. The invention of the automobile and aircraft shortly thereafter served to accelerate the design methodology for structural components.

The design process may be described concisely as the procedure whereby the loads, geometry, and material properties to be used in a given structural component are chosen so as to ensure that all design constraints are met. Design constraints can be numerous. Indeed, they may be so numerous and onerous as to make a successful design essentially impossible to build on this planet. An example of this fact is the so-called National Aerospace Plane (NASP), proposed during the Reagan presidency in the 1980s, as shown in Fig. 7.8. The aircraft was intended to be a single-stage-to-orbit vehicle, which would require velocities in the vicinity of Mach 20. This aircraft was not possible to build within the technology available at that time because there were no materials available that were capable of simultaneously meeting the aerodynamic heating and weight constraints required to make

Fig. 7.9 Photo of base support system for a highway sign designed to fail if impacted by a vehicle

the aircraft cost effective. Thus, it should be recognized that there are very real circumstances wherein the design constraints can preclude an acceptable design.

The essential nature of the design process is to invert the algebraic equations resulting from the solution of the governing differential equations comprising the model to produce a form of the equations that treats the input loads, geometry, and material properties as outputs in terms of the stresses, strains, and displacements. The ingenious designer can then optimize the inputs for whatever purpose he or she may have in mind. Of course, this process assumes tacitly that the designer has adequate information at hand to know what the design constraints are, and these are not always known concisely.

In this course, we will avoid much of the complexity associated with design constraints in order to illustrate the design process in a straightforward way that is at once demonstrative and informative without being excessively cumbersome.

An example of a modern structure that is simple but effective is shown in Fig. 7.9, which shows the base of a highway sign designed to withstand axial and bending stresses introduced by wind loading, and also designed to fail if struck by a vehicle, thus improving highway safety.

In order to illustrate the design process, we will confine our attentions to the design of structural components that are long and slender (we call them bars). Fortunately for us, we have already introduced three different models for analyzing bars: uniaxial bars, torsion bars, and beams. Before considering the design process further, we will first explore a procedure for analyzing bars that are subjected to combined loadings, as described in the next section.

7.2 A Procedure for Analyzing Bars Subjected to Combined Loading

We will show in this section that when a bar is subjected to loadings that cause extension, torsion, and bending simultaneously it may be analyzed using the models already developed in Chaps. 3, 4, and 5. The procedure that will be utilized is called *superposition.* The principle of superposition for analysis of structural bars is stated as follows: ***given a bar of specified geometry and material makeup, the response of the bar to a combination of axial, torsion, and bending loads may be obtained by superposing the response of the bar to each of the loads separately.***

The above concept is quite straightforward, and its usage will result in a very powerful design tool, as will be seen shortly. In order to demonstrate the power of this principle, consider the example shown in Fig. 7.10.

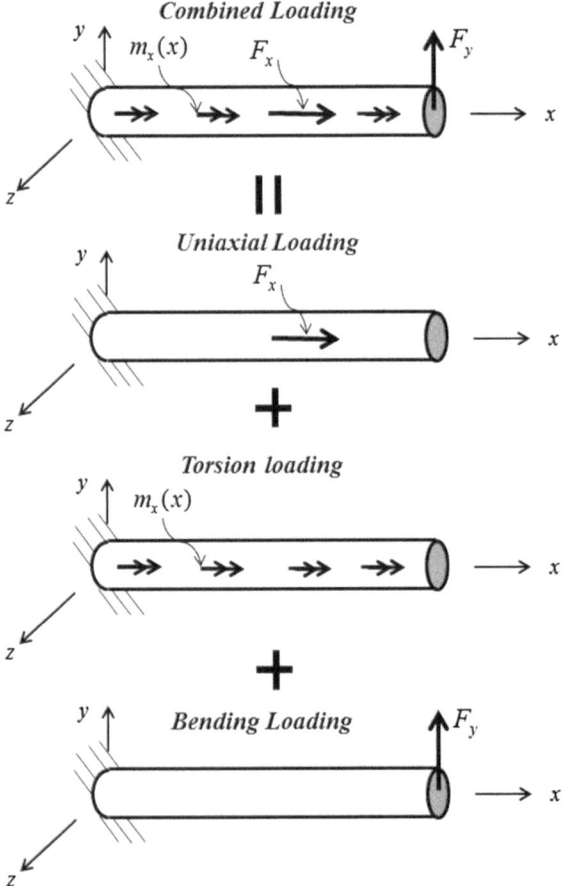

Fig. 7.10 Illustration of the principle of superposition

The above principle can be proven to be a sufficient condition for application to uniaxial bars by introducing the mathematical definition of linearity. Linearity of an equation is defined in the following way. Given an equation of the form (Reddy 1984)

$$L[u(x)] = f(x) \tag{7.1}$$

where L is an operator acting on the dependent variable, u, and x is the independent variable, the equation is defined to be linear if and only if the following two conditions are met

$$\text{1. Homogeneity: } L[\alpha u] = \alpha L[u] \tag{7.2a}$$

$$\text{2. Superposition: } L[u_1 + u_2] = L[u_1] + L[u_2] \tag{7.2b}$$

where α is an arbitrary constant.

From the above conditions, it can be seen that whenever an equation (or set of equations) is linear, the principle of superposition may be employed as described above. Although the principle of homogeneity is not as important, it is nevertheless useful. In vernacular, it may be stated as follows: *if the input load is increased by a factor k, then the output stresses, strains, and displacements are increased by the factor k.* Furthermore, it can be shown using (7.2) that **each and every equation used in the models developed herein for bars is linear!**

Example Problem 7.1
Given: The equation describing the axial displacement in a uniaxial bar is given by

$$\frac{d}{dx}\left(EA\frac{du}{dx}\right) = -p_x(x)$$

Required: Show that the above equation is linear.

Solution: First write the equation in the form of (7.1) such that

$$L[u(x)] = f(x) \tag{E.7.1.1}$$

where

$$L[u] \equiv \frac{d}{dx}\left(AE\frac{du}{dx}\right) \tag{E.7.1.2}$$

and

$$f(x) \equiv -p_x(x) \tag{E.7.1.3}$$

Now check for homogeneity as follows

$$L\left[\alpha u\right] = \frac{d}{dx}\left(AE\frac{d(\alpha u)}{dx}\right) = \alpha\frac{d}{dx}\left(AE\frac{du}{dx}\right) = \alpha L\left[u\right] \qquad (E.7.1.4)$$

due to the commutative property of differentiation. Now check for superposition as follows

$$L\left[u_1 + u_2\right] = \frac{d}{dx}\left(AE\frac{d(u_1 + u_2)}{dx}\right) = \frac{d}{dx}\left(AE\frac{du_1}{dx}\right) + \frac{d}{dx}\left(AE\frac{du_2}{dx}\right)$$
$$= L\left[u_1\right] + L\left[u_2\right] \qquad (E.7.1.5)$$

Therefore the homogeneity and superposition properties are satisfied and the differential equation is linear. Note that the geometry and material properties are included in the operator $L[.]$, so that these must be the same when superposition is used, but the loads, as represented by the forcing function $f(x)$ may be superposed.

Example Problem 7.2
Given: The highway sign shown below is subjected to the loading shown.

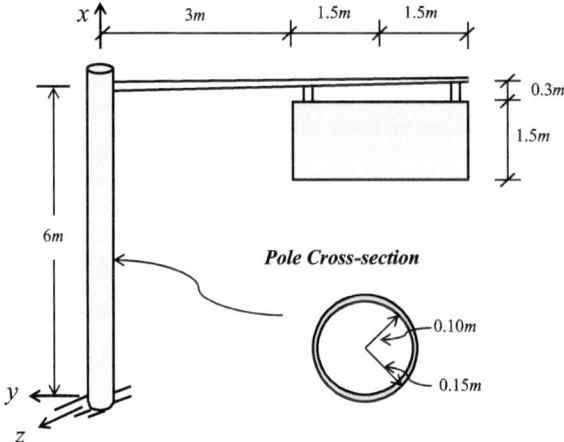

Assumptions

1. The maximum wind blows normal to the sign and produces an evenly distributed pressure of 0.01 MPa.
2. The vertical member is made of steel that has a weight per unit volume of 7,850 kg/m^3.
3. The weight of the horizontal member is included in the sign weight and therefore may be neglected.
4. The sign is homogenous and weighs 500 kg.

Required

1. Resolve the loads to the vertical member and draw a free body diagram of the vertical member with the x-axis (long axis) constructed horizontally (you will need to rotate the vertical member 90 clockwise).
2. Draw four separate depictions of the vertical member, showing the uniaxial, torsion, and bending loads (in two planes) constructed separately.
3. Using results obtained in previous homework or example problems, construct the state of stress in the uniaxial, torsion, and bending problems at the point $(x = 0, \; y = 0, \; z = 0.15\,\text{m})$.
4. Superpose the stress states obtained for the uniaxial, torsion, and bending problems to depict the state of stress on the vertical member at the point $(x = 0, \; y = 0, \; z = 0.15\,\text{m})$.
5. Draw Mohr's circles for the state of stress at the point $(x = 0, \; y = 0, \; z = 0.15\,\text{m})$ and determine the principal stresses and maximum shear stress.
6. Check for failure at the point $(x = 0, \; y = 0, \; z = 0.15\,\text{m})$ assuming

$$\sigma^{\mathrm{T}} = 275\,\text{MPa}, \; \sigma^{\mathrm{C}} = 275\,\text{MPa}, \; \sigma^{\mathrm{S}} = 125\,\text{MPa}$$

Solution

1. The weight per unit length of the pole is first determined by calculating the cross-sectional area of the pole as follows:

$$A = \pi(r_o^2 - r_i^2) = \pi(0.15^2 - 0.10^2) = 0.03927\,\text{m}^2 \qquad (\text{E}.7.2.1)$$

The force per unit length applied axially to the bar is therefore obtained by multiplying the weight per unit volume of the bar by the cross-sectional area and the gravitational constant as follows:

$$p_x = -7850\,\frac{\text{kg}}{\text{m}^3} \times 0.03927\,\text{m}^2 \times 9.80665\,\frac{\text{N}}{\text{kg}} = 3023\,\frac{\text{N}}{\text{m}} \qquad (\text{E}.7.2.2)$$

Next, the force caused by the weight of the sign is as follows:

$$F_x = 500\,\text{kg} \times 9.80665\,\frac{\text{N}}{\text{kg}} = 4903.3\,\text{N} \qquad (\text{E}.7.2.3)$$

Furthermore, it can be seen from the diagram above that the net force caused by the wind loading is as follows:

$$F_z = 0.01\,\text{MPa} \times \frac{10^6\,\text{N/m}^2}{\text{MPa}} \times 3\,\text{m} \times 1.5\,\text{m} = 0.045 \times 10^6\,\text{N} \qquad (\text{E}.7.2.4)$$

Thus the resulting loads on the structure are as shown below

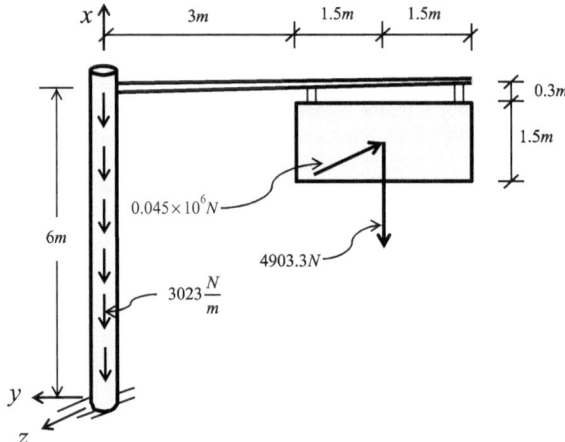

Next, the loads are resolved to the end of the horizontal member using the following free body diagram

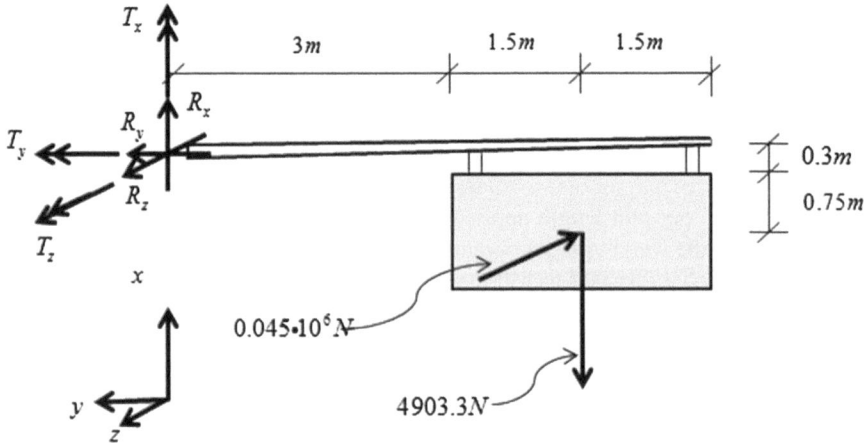

Summing forces to obtain the reactions in the above free body diagram gives the following, where all units are in N:

$$\sum F_x = 0 = R_x - 4903.3 \Rightarrow R_x = 4903.3 \qquad \text{(E.7.2.5)}$$

$$\sum F_y = 0 = R_y \qquad \text{(E.7.2.6)}$$

$$\sum F_y = 0 = R_z - 0.045 \times 10^6 \Rightarrow R_z = 0.045 \times 10^6 \qquad \text{(E.7.2.7)}$$

Similarly, summing moments about the point where the reactions intersect in the above diagram gives the following, where all units are in Nm:

$$\sum M_x = 0 = T_x + 0.045 \times 10^6 \times 4.5 \Rightarrow T_x = -0.2025 \times 10^6 \qquad \text{(E.7.2.8)}$$

$$\sum M_y = 0 = T_y - 0.045 \times 10^6 \times 1.05 \Rightarrow T_y = 0.04725 \times 10^6 \qquad \text{(E.7.2.9)}$$

$$\sum M_z = 0 = T_z - 4903.3 \times 4.5 \Rightarrow T_z = 22,065 \qquad \text{(E.7.2.10)}$$

Thus, rotating the pole clockwise and noting that the reactions on the horizontal are equal in magnitude and opposite in sign to the loads applied to the pole results in the following free body diagram for the pole

Combined Loading Diagram

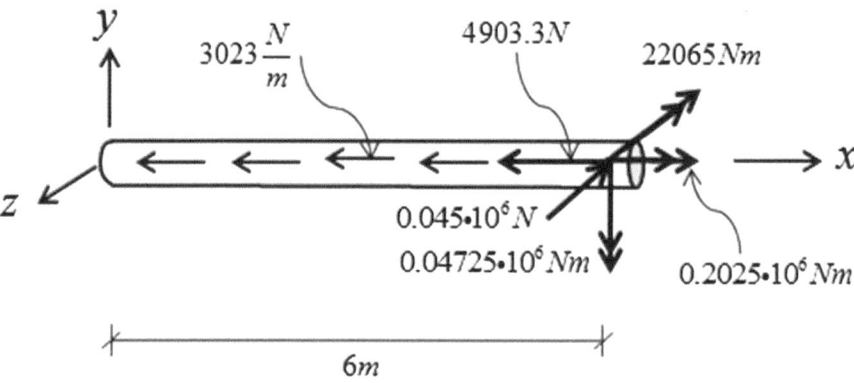

2. The above diagram may in turn be resolved into the following four load diagrams.

Uniaxial Loading Diagram

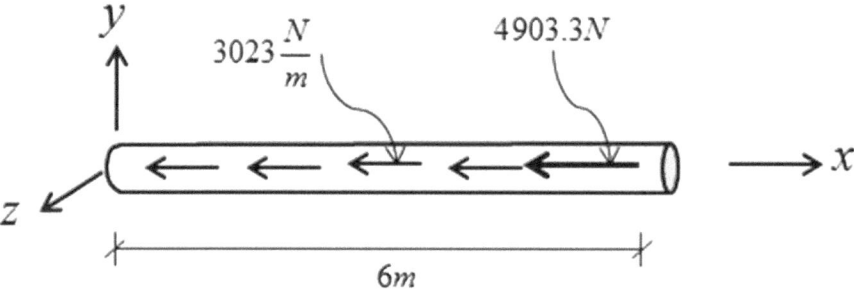

Torsion Loading Diagram

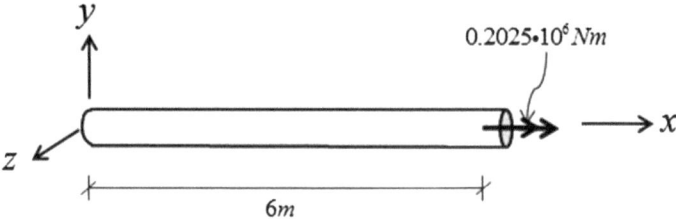

x-y Plane Bending Load Diagram

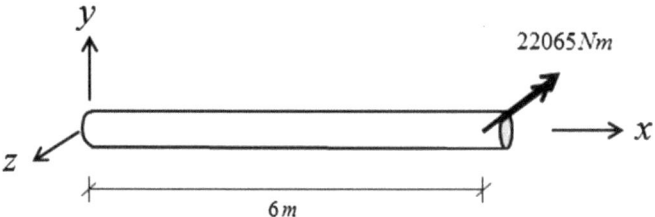

x-z Plane Bending Load Diagram

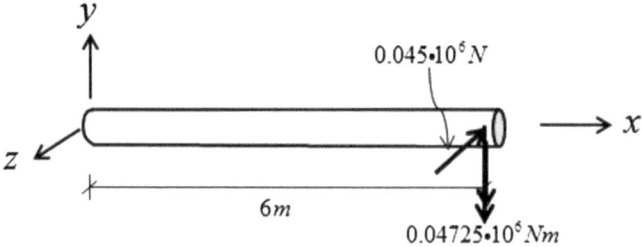

3. Part (a)—Uniaxial Loading Diagram: By summing forces in the x direction we know that the internal force at $x = 0$ is as follows:

$$\sum F_x = 0 = P\,(x = 0) - 4{,}903.3 - 3{,}203 \times 6 \Rightarrow P\,(x = 0)$$

$$= -23{,}041\,\text{N} \tag{E.7.2.11}$$

From Chap. 3, we know that for uniaxial bars

$$\sigma_{xx}^{U}(x = 0) = \frac{P(x = 0)}{A} = \frac{-23041\,\text{N}}{0.03927\,\text{m}^2} \Rightarrow \sigma_{xx}^{U}(x = 0) = -0.5867\,\text{MPa} \tag{E.7.2.12}$$

Part (b)—Torsion Loading Diagram: Summing moments about the x-axis at $x = 0$, we know that

$$\sum M_x = 0 = -M_x(x = 0) + 0.2025 \times 10^6 \, \text{Nm} \Rightarrow M_x(x = 0)$$
$$= 0.2025 \times 10^6 \, \text{Nm} \tag{E.7.2.13}$$

The polar moment of inertia, J, is given by

$$J = \frac{\pi}{2}(r_0^4 - r_i^4) = \frac{\pi}{2}(0.15^4 - 0.10^4) \Rightarrow J = 0.00,040,625 \, \text{m}^4 \tag{E.7.2.14}$$

From Chap. 4, we know that for torsion bars

$$\sigma_{x\theta}^T(x = 0, \ r = r_o) = \frac{M_x(x = 0) \, r_o}{J} = \frac{0.2025 \times 10^6 \, \text{Nm} \times 0.15 \, \text{m}}{0.00040625 \, \text{m}^4}$$
$$\Rightarrow \sigma_{x\theta}^T(x = 0, \ r_0 = z = 0.15) = 74.78 \, \text{MPa} \tag{E.7.2.15}$$

Part (c)—$x - y$ Plane Bending Load Diagram: from Chap. 5 we know that at the coordinate location $y = 0$ there is no axial stress in the beam due to bending in the $x - y$ plane.

Part (d)—$x - z$ Plane Bending Load Diagram: By summing moments about the y-axis at $x = 0$, we know that

$$\sum M_y = 0 = -M_y(x = 0) + 0.045 \times 10^6 \, \text{N} \times 6 \, \text{m} - 0.04725 \, \text{Nm}$$
$$\Rightarrow M_y(x = 0) = 0.22275 \times 10^6 \, \text{Nm} \tag{E.7.2.16}$$

The moment of inertia about the y-axis is given by

$$I_{yy} = \frac{\pi}{4}(r_o^4 - r_i^4) = \frac{\pi}{4}(0.15^4 - 0.10^4) \Rightarrow I_{yy} = 0.000203125 \, \text{m}^4 \tag{E.7.2.17}$$

From Chap. 5 we know that for beams in bending (in the x–z plane there is a sign change in the formula)

$$\sigma_{xx}^B(x = 0, z = 0.15) = \frac{M_y(x = 0) \, z}{I_{yy}} = \frac{0.022275 \times 10^6 \, \text{Nm} \times 0.15 \, \text{m}}{0.000203125 \, \text{m}^4}$$
$$\Rightarrow \sigma_{xx}^B(x = 0, z = 0.15) = 164.6 \, \text{MPa} \tag{E.7.2.18}$$

Thus, the states of stress at the coordinate location for the four loading diagrams are as follows:

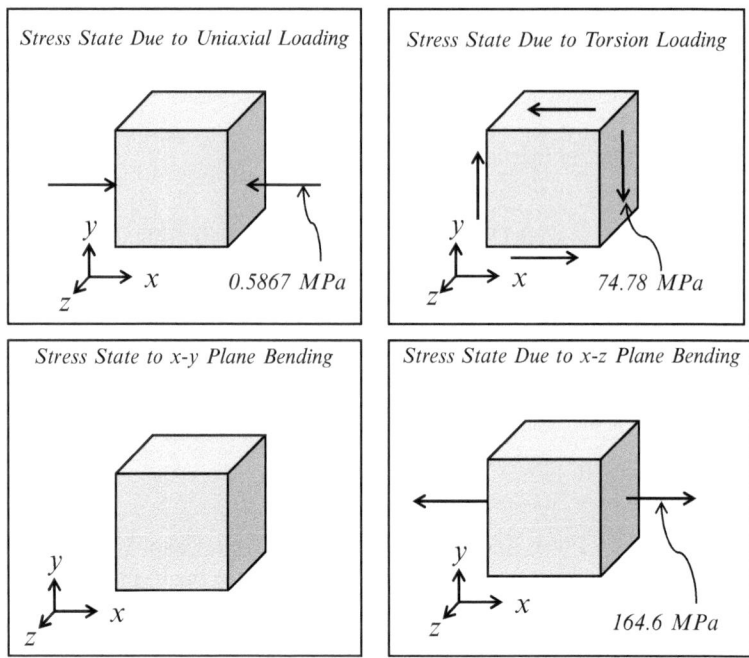

4. Superposing the above four states of stress at the point $(x = 0, y = 0, z = 0.15\,\text{m})$ results in the following state of stress due to all four loadings applied to the pole

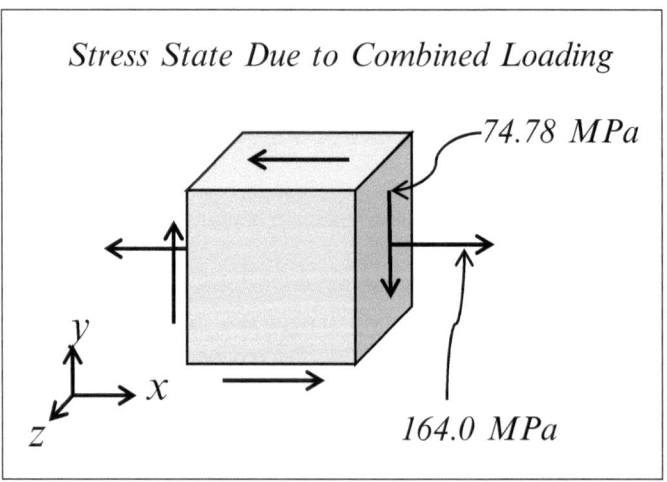

5. Mohr's circles for the point $(x = 0, y = 0, z = 0.15\,\text{m})$ are as follows:

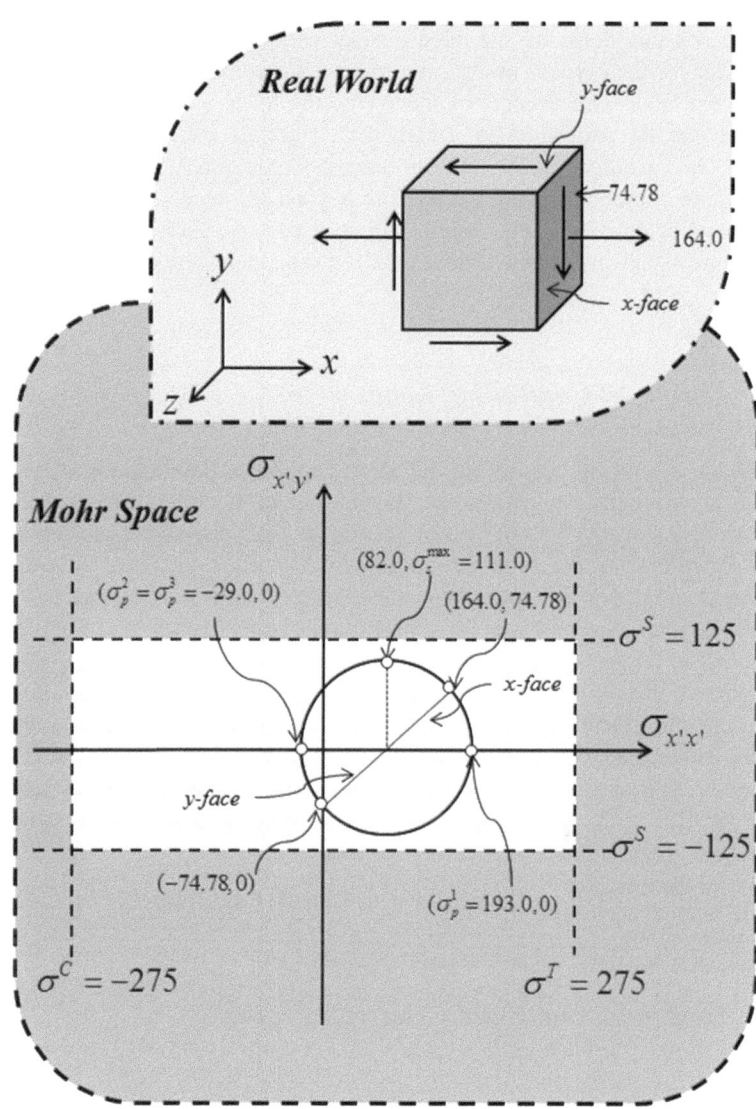

6. Since the circle falls inside the failure envelope in the above diagram it is clear that it is predicted that the point $(x = 0,\ y = 0,\ z = 0.15\,\text{m})$ does not fail.

7.3   The Design Process

In practice, the design process involves solving what mathematicians typically call an "inverse" problem. Essentially, the output stresses, strains, and displacements are treated as if they are known because the design process requires that one or more of the outputs not exceed some specified critical value. The model is then prescribed as if the loads, geometry, and material properties are unknowns for the simple reason that these are the three distinct variables that can be modified by the designer for the purpose of ensuring that the structural design meets all of the design constraints. This process of modifying the loads, geometry, and material properties can be performed in a somewhat rigorous methodology such as a structural optimization algorithm. For complex structures, this approach can become quite complicated, and is therefore beyond the scope of the present text. For purposes of illustrating the design process in a simple yet ingenious way, each of the controllable variables will be considered separately herein.

7.3.1   Design by Controlling Loads

Altering the design by controlling the loads applied to the object is in practice normally the simplest design change possible. In most circumstances designers utilize models that involve only linear equations. Thus, the applied loads may be simply scaled down if the initial design process results in a predicted failure of the object in question. For example, if the stresses on a forklift are expected to exceed the allowable stresses by 25 % for a maximum design loading of 10,000 N, the allowable load for use with the forklift could simply be reduced by 20 % to 8,000 N. This approach may sound simple minded, but it is not far-fetched, especially in the case where the object has already been constructed. An example wherein this occurred is the case of the extremely large military transport aircraft C5A Galaxy, first introduced in the late 1960s, as shown in Fig. 7.11. This aircraft was designed to transport large vehicles such as tanks and other aircraft (really!). Unfortunately, design flaws required that the allowable loads be reduced below that initially planned for the aircraft.

7.3.2   Design by Controlling Material Properties

Sometimes it is necessary to employ changes in materials in order to produce a design that passes all of the design constraints. An obvious example is the case of a structural component that is to be subjected to extremely high temperatures that would be sufficient to melt one material but not another. This circumstance is common in hot gas turbine engines, wherein temperatures can easily exceed the melting temperature of aluminum. In this case, high temperature nickel-based alloys are typically used, such as Inconel.

Controlling the material used in a particular design scenario often comes down to a matter of cost, since different materials have different unit costs. For this purpose, it is common to compare the modulus of elasticity of materials per cost per kg.

Fig. 7.11 C5A Galaxy military aircraft (photo courtesy USAF)

Since the mass density of the material may also be an important factor (meaning loads!) in the design, it is also common to compare cost per mass density of various materials that may be used in the design process.

The process of choosing materials for a particular design can become quite complicated when cost is included as a design constraint (and it always is). For example, consider the process of designing a roadway made of a mixture of cementitious and asphaltic pavement. Both of these types of pavement contain a mastic (cement or asphalt), an aggregate (rocks), reinforcement (steel), and additives. The choice of the material to be used for each of these components in the mix will depend on the availability and cost of the various choices, and each will affect the overall performance of the resulting roadway. Indeed, it is not uncommon for a contractor to find at least four or five different aggregates within a few hundred km of the construction site.

Thus, it can be seen that, depending on the application, controlling the selection of materials for the purpose of meeting all of the design constraints can be a rather complicated process.

7.3.3 Design by Controlling Geometry

We are approaching the end of this text, but we have saved some of the most interesting material for the very last. The final variable that the designer has the ability to control in the design process is geometry. This is by far the most open-ended, artistically interesting variable that can be controlled in the design process. We have developed models in this text that will provide constraints on the feasibility of a design. These models should be utilized with care; that is, they should not be violated without good cause. Still, for every application, there are infinite numbers of geometric configurations that will satisfy all of the design constraints. This then is where the practicing engineer will have the opportunity

Fig. 7.12 Photos of the author with a giant redwood in the Mariposa Grove at Yosemite National Park

to choose aesthetically pleasing possibilities, just as Imhotep did when he built the first step pyramid for Djoser, or when the first flying buttress was applied to a Gothic cathedral, or when Gustav Eiffel designed his fabulous tower. And what a design it is—can there be any doubt that the Eiffel Tower is the most recognizable structure on Earth?

The reader is challenged to use his/her ingenuity to design at the leading edge of what is humanly possible. In the most successful circumstance, the reader may well even approach the design provided by nature herself. For example, consider the case of a giant redwood tree, as shown in Fig. 7.12. What better example can be made for ingenious and yet attractive design? Note that through millions of years of evolution, this species of tree has managed to reach to heights in excess of 100 m by evolving its own moment of inertia as a function of height in such a way as to reduce stresses [see (5.21)] to a level that is within the failure envelope for the tree, so that structural failure is obviated. Indeed, most trees of this species *die standing*—structural failure occurs rarely in giant redwoods.

Thus, armed with the analytic tools developed by our forefathers, and reviewed in this textbook, the enterprising designer can go forth and, if he or she is fortunate, follow in the footsteps of Gustav Eiffel.

Example Problem 7.3

Given: A structural designer is given the following challenge—to design a 30 m tall flagpole to fly a flag that is to display a flag that is 6 m by 12 m at the top of the mast.

Required: Design a flagpole that will not fail structurally due to yielding or fracture.

Solution: A careful examination of a flagpole will reveal that it is a very simple structure, in the sense that in order to make it easy to raise and lower the flag, it is normally attached to a cord that is attached to the top of the mast and hoisted by hand. This means that the only point at which load is applied to the flag (other than its own weight) is at the top of the mast. Furthermore, due to the load carrying behavior of rope (it can only carry uniaxial load), the loading caused by the flag on the pole acts through the centroid of the mast. Thus, torsion can be induced by the flag itself.

It will be assumed that the maximum wind speed encountered where the flagpole is to be constructed is 170 km/h (approximately 100 mi/h). Experiments show that the load applied to the pole is proportional to the area of the flag and the velocity squared. Furthermore, for a flag of area 1 m^2 subjected to a velocity of 50 km/h, experiments indicate that the force applied to the flag is approximately 100 N. For purposes of demonstration, it is assumed that the wind loading on the flagpole itself may be neglected. Thus, for the flag described above, the total force applied to the pole (by the flag) when the wind velocity is a maximum is approximately

$$F = 100\,\text{N} \times \frac{6\,\text{m} \times 12\,\text{m}}{1\,\text{m}^2} \times \left(\frac{170\,\text{km/h}}{50\,\text{km/h}}\right)^2 \Rightarrow F = 83.23 \times 10^3\,\text{N} \qquad \text{(E.7.3.1)}$$

Thus, the flagpole is represented by the following diagram.

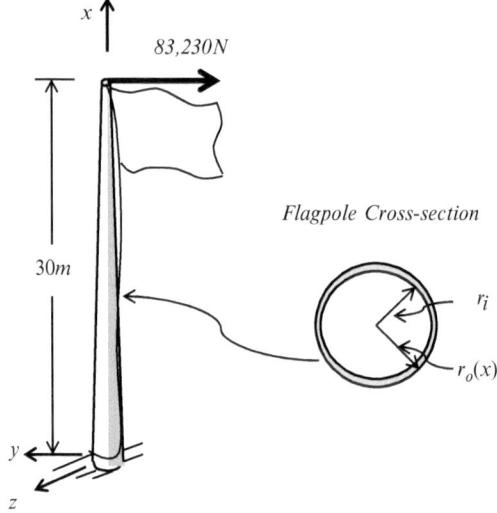

It can be seen from Example Problem 5.2 that the vertical component of the normal stress due to the above loading will be given by

$$\sigma_{xx}^{\text{bending}} = -\frac{F(x-L)\,y}{I_{zz}} = -\frac{83.23 \times 10^3 \, \text{N}(x-30)\,y}{I_{zz}} \qquad \text{(E.7.3.2)}$$

All other components of stress in the pole are negligible. However, the weight of the pole itself may produce significant normal stress in the vertical direction, and from Homework Problem 3.3 this is given by

$$\sigma_{xx}^{\text{uniaxial}} = \frac{P}{A} = \frac{\rho g^E A(x-L)}{A} = 9.8066 \times 10^{-3}\rho(x-30)\,\text{N/g} \qquad \text{(E.7.3.3)}$$

Thus, superposing the two stresses above results in the following:

$$\sigma_{xx} = -\frac{83.23 \times 10^3 \text{N}(x-30)y}{I_{zz}} + 9.8066 \times 10^{-3}\rho(x-30)\,\text{N/g} \qquad \text{(E.7.3.4)}$$

It is assumed herein that the only design constraint is failure due to excessive stresses. Normally at this point it would be prudent to draw Mohr's circle, but in the present problem the stress state at the critical points (along the external surface of the flagpole) is everywhere uniaxial, so that the normal stress in the x direction is a principal stress. Thus, the above equation can be used as the design constraint by setting a maximum allowable value of the stress component σ_{xx} equivalent to σ^C, as the maximum magnitude of normal stress (depending on how the cross-section is shaped) occurs in compression.

It will be assumed herein that the flagpole is to be constructed from A36 steel, with properties as described in the Appendix. A36 steel has a density of $\rho = 7.8\,\text{mg/m}^3$ and a compressive strength of $\sigma^C = 250\,\text{MPa}$. Furthermore, due to the fact that the wind can blow in any direction, it will be necessary for the pole to be circular in cross-section. It will be assumed for manufacturing purposes that the inner radius of the flagpole remains constant along the length. This will produce a moment of inertia given by

$$I_{zz} = \frac{\pi}{4}\left[r_o(x)^4 - r_i^4 \right] \tag{E.7.3.5}$$

where r_i is the inner radius and $r_o(x)$ is the outer radius, as shown in the figure above. Substituting this formula and the material properties for A36 steel into (E.7.3.4) results in the following:

$$-250 \times 10^6\,\text{N/m}^2 \le -\frac{83.23 \times 10^3\,\text{N}\,(x - 30)\,r_o(x)}{\frac{\pi}{4}\left[r_o(x)^4 - r_i^4 \right]} + 9.8066 \times 10^{-3} \times 7.8 \times 10^6 (x - 30)\,\text{Pa} \tag{E.7.3.6}$$

The above is a nonlinear equation in $r_o(x)$. As such, it may prove quite difficult to obtain an optimum design. For purposes of the current design, it will be assumed that the maximum stress occurs at the end $x = 0$. Furthermore, an inner radius, $r_i = 0.2\,\text{m}$ will be chosen. Also, because the loading goes to zero at the top of the flagpole, the cross-sectional area may also go to zero from a structural standpoint. Thus, for manufacturing reasons, the outer radius at the top of the flagpole will be set to $r_o(x = 30) = 0.21\,\text{m}$. Furthermore, for manufacturing reasons it will be assumed that the outer radius of the flagpole is linear in x. Using the above information the description of the outer radius as a function of x is given by

$$r_o(x) = r_o(0) - (r_o(0) - 0.21)\frac{x}{30} \tag{E.7.3.7}$$

Substituting the above into (E.7.3.6) results in the following equation:

$$-250 \times 10^6\,\text{N/m}^2 \le -\frac{83.23 \times 10^3\,\text{N}\,(x - 30)\left\{ r_o(0) - (r_o(0) - 0.21)\frac{x}{30} \right\}}{\frac{\pi}{4}\left[\left\{ r_o(0) - (r_o(0) - 0.21)\frac{x}{30} \right\}^4 - 0.0016 \right]} + 9.8066 \times 10^{-3} \times 7.8 \times 10^6 (x - 30)\,\text{Pa} \tag{E.7.3.8}$$

The above equation is a nonlinear equation in x and $r_o(0)$. As such, it may be very difficult to solve. Suppose instead that it is assumed that the most important location to consider is at $x = 0$. In this case the above equation simplifies to the following:

$$-250 \times 10^6 \, \text{N/m}^2 \le -\frac{83.23 \times 10^3 \text{N} \times 30\,\text{m} \times r_{\text{o}}(0)}{\frac{\pi}{4}\left[r_{\text{o}}(0)^4 - 0.2^4\right]}$$
$$- 9.8066 \times 10^{-3} \times 7.8 \times 10^6 \times 30\,\text{Pa} \tag{E.7.3.9}$$

The above equation can be rearranged to give the following:

$$250 \ge \frac{3.179\, r_{\text{o}}(0)}{\left[r_{\text{o}}(0)^4 - 0.0016\right]} + 2.295 \tag{E.7.3.10}$$

where $r_{\text{o}}(0)$ is in units of meters. Evaluating the above inequality for successively increasing values of $r_{\text{o}}(0)$ will result in a minimum allowable outer radius at the base of the flagpole given by

$$\boxed{r_{\text{o}}^{\min}(0) = 0.2662\,\text{m}} \tag{E.7.3.11}$$

It would appear that our design is now complete, having chosen the loads, geometry, and material properties. However, the discerning reader may well realize that the geometry proposed above may induce failure at some other point along the long axis of the flagpole. In order to check for this possibility, (E7.3.8) should be checked for values of x greater than zero. This question is perhaps best answered with the aid of a mathematics software package. The results of such an exercise, shown below, indicate that while the stress does increase slightly in the first two feet of the flagpole from the base, the stress does in fact remain below the critical value of 250 MPa. Therefore, the design is complete.

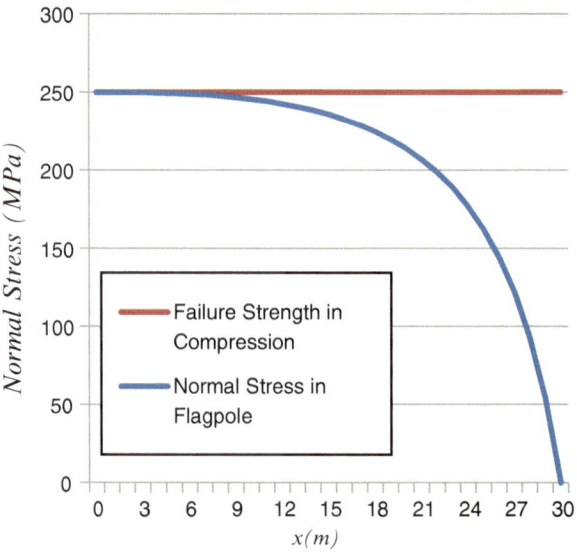

The final design of the flagpole is shown below.

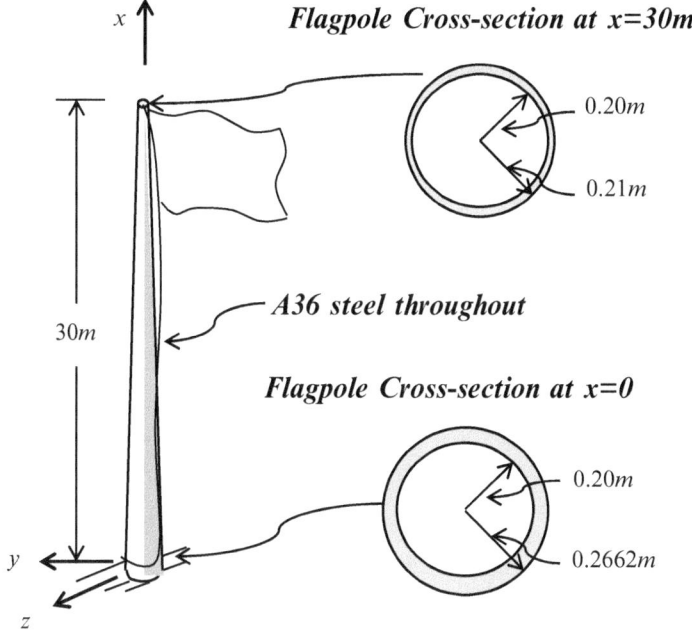

7.4 Assignments

PROBLEM 7.1

GIVEN: From Chap. 5 we know that the equation governing the transverse displacement of a beam is given by

$$\frac{d^2}{dx^2}\left[EI_{zz}\frac{d^2 v_0}{dx^2}\right] = p_y(x)$$

REQUIRED

1. Show that the above equation satisfies superposition and homogeneity and is therefore linear.

PROBLEM 7.2

GIVEN: The structure below is subjected to loading as shown.

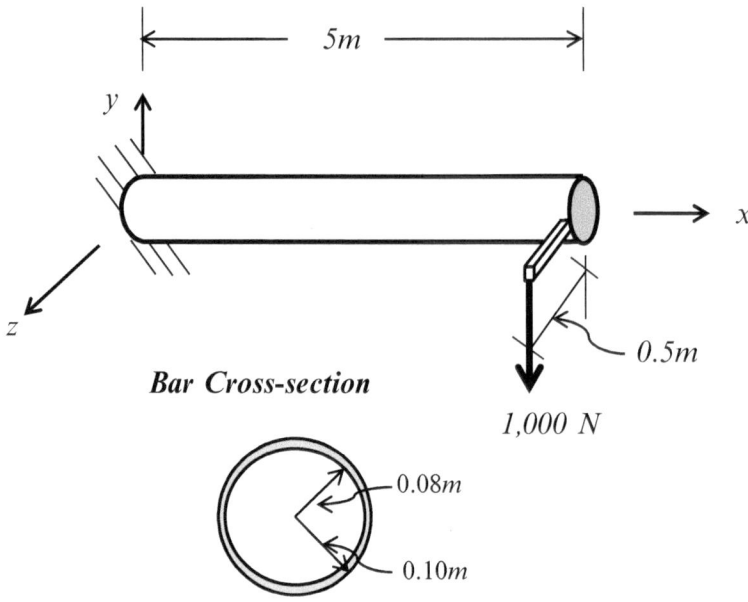

Bar Cross-section

ASSUMPTIONS

1. The weight of the bar may be neglected.
2. The failure strength is

$$\sigma^{\mathrm{T}} = 250\,\mathrm{MPa}, \ \sigma^{\mathrm{C}} = 250\,\mathrm{MPa}, \ \sigma^{\mathrm{S}} = 100\,\mathrm{MPa}$$

REQUIRED

1. Resolve the load to the centroid of the bar and draw the resulting depiction.
2. Draw separate depictions of the bar showing the separate loads.
3. Using previous results obtained in the course, construct the stress state at the coordinate location $(x = 0, \ y = 0.10\,\mathrm{m}, \ z = 0)$.
4. Superpose the stress states obtained in (3) above and draw the resulting stress state.
5. Draw Mohr's circles for the state of stress obtained in (4).
6. Check for failure.

PROBLEM 7.3

GIVEN: The structure below is subjected to loading as shown.

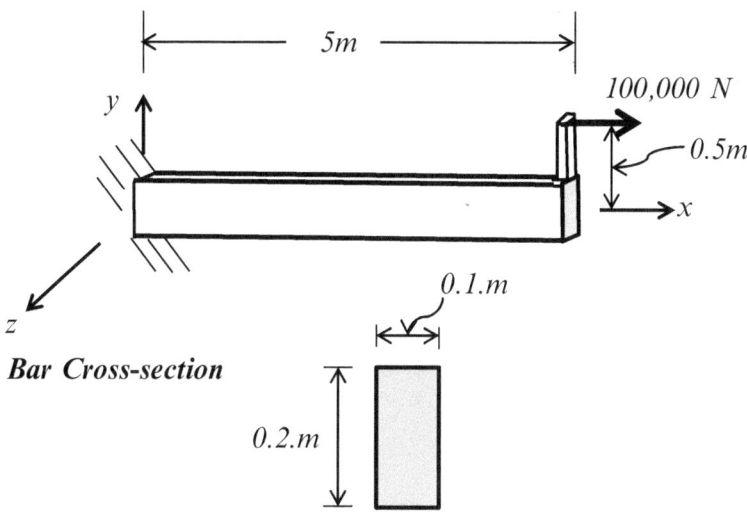

ASSUMPTIONS

1. The weight of the bar may be neglected.
2. The failure strength is $\sigma^T = 200\,\text{MPa}$, $\sigma^C = 200\,\text{MPa}$, $\sigma^S = 70\,\text{MPa}$.

REQUIRED

1. Resolve the load to the centroid of the bar and draw the resulting depiction.
2. Draw separate depictions of the bar showing the separate loads.
3. Using previous results obtained in the course, construct the stress state at the coordinate location $(x = 0,\ y = 0.10\,\text{m},\ z = 0)$.
4. Superpose the stress states obtained in (3) above and draw the resulting stress state.
5. Draw Mohr's circles for the state of stress obtained in (4).
6. Check for failure.

PROBLEM 7.4

GIVEN: The parallel bars shown below are 3 m long and 1.5 m tall and are intended to carry the weight of a gymnast.

Photo of Norwegian gymnast Espen Jansen performing at the Norwegian National Championships in 2001

REQUIRED: Design the horizontal and vertical members exclusive of the base of the structure.

PROBLEM 7.5
GIVEN: Patients with leg injuries are often required to use crutches.
REQUIRED: Design a pair of crutches that will satisfy the following design constraints:

1. Must be lightweight.
2. Must be collapsible to a length of less than 0.8 m.
3. Cost effective.
4. Attractive.
5. Ergonomic.
6. Capable of carrying the weight of a person weighing 125 kg.

PROBLEM 7.6
GIVEN: The camera stand shown below is intended to hold a camera weighing up to 10 kg at a height of 1.5 m. In addition, the stand should be collapsible into a size that will fit into a piece of luggage that is 0.5 m long, as shown.

REQUIRED: Design the structural support system.

PROBLEM 7.7
GIVEN: The shopping baskets shown below are designed to carry up to 30 kg in merchandise.

REQUIRED: Design the handle for the shopping basket.

PROBLEM 7.8
GIVEN: The stool shown below is designed to carry the weight of a person.

REQUIRED: Design the structural support system.

PROBLEM 7.9
GIVEN: The wheelchair shown below.

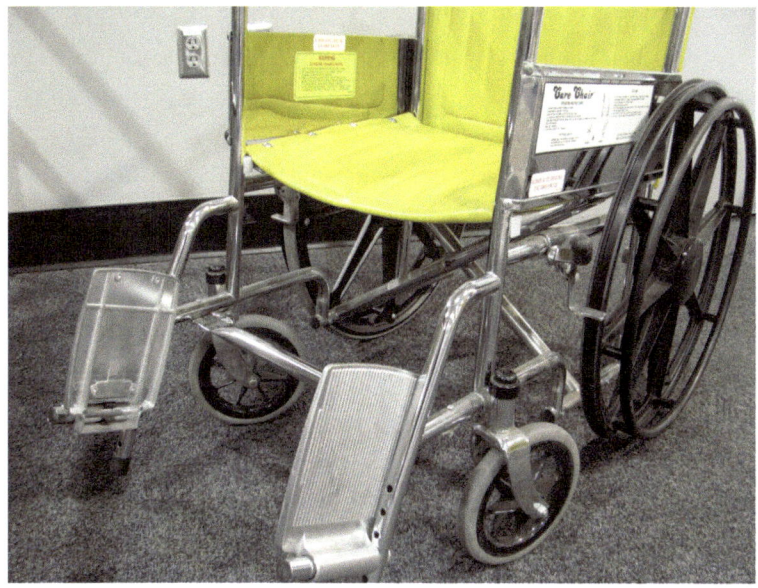

REQUIRED: Design an effective support system to carry the passenger's feet.

References

Reddy J (1984) An Introduction to the Finite Element Method. McGraw-Hill, New York

Appendix A
Mechanical properties of structural materials

Material class	Alloy or grade	Density (mg/m^3)	E (GPa)	G (GPa)	v	σ^T (MPa)	σ^C (MPa)	σ^S (MPa)
Steel	A36	7.8	200	75	0.26	250	250	100
	Stainless 304	8.0	193	70	0.28	205	205	74
Aluminum	6061-T6	2.7	68.9	26	0.33	276	276	184
	2014-T6	2.8	73	28	0.33	414	414	225
Magnesium	Am 1004-T61	1.83	45	18	0.35	152	152	–
Titanium	Ti-6Al-4V	4.51	110	41	0.34	830	830	–
Cast Iron	Malleable (ASTM 47)						52	
	Grey (ASTM 20)	7.19	67	27	0.28	50	50	–
Copper	99.9% pure	8.92	110	48	0.34	62	62	–
Concrete	Low strength	2.38	22	–	0.15	–	14	12
	High strength	2.38	29	–	0.15	–	34	38
Wood	Red Oak	0.6–0.9	12	4.65	0.29	5.1	5.0	–
	Douglas Fir	0.47	11–13		0.29			–

D.H. Allen, *Introduction to the Mechanics of Deformable Solids: Bars and Beams*, 213
DOI 10.1007/978-1-4614-4003-1, © Springer Science+Business Media New York 2013

Author Index

Subject Index